# Mobile Communication Satellites

EHSAN TALEBI
(415) 941- 9226

## Other McGraw-Hill Communications Books of Interest

*To order or receive additional information on these or any other McGraw-Hill titles, in the United States please call 1-800-822-8158. In other countries, contact your local McGraw-Hill representative.*

**BC14BCZ**

# Mobile Communication Satellites

Tom Logsdon

**McGraw-Hill, Inc.**

New York   San Francisco   Washington, D.C.   Auckland   Bogotá
Caracas   Lisbon   London   Madrid   Mexico City   Milan
Montreal   New Delhi   San Juan   Singapore
Sydney   Tokyo   Toronto

**Library of Congress Cataloging-in-Publication Data**

Logsdon, Tom (date).
    Mobile communication satellites / Tom Logsdon.
        p.    cm.
    Includes bibliographical references and index.
    ISBN 0-07-038476-2 (alk. paper)
    1. Artificial satellites in telecommunication.    2. Mobile
communication systems.    I. Title.
TK5104.L64    1995
384.5'3—dc20                                    94-23063
                                                CIP

1 2 3 4 5 6 7 8 9 0    DOC/DOC    9 0 0 9 8 7 6 5

ISBN 0-07-038476-2

*The sponsoring editor for this book was Stephen S. Chapman, the editing
supervisor was Caroline R. Levine, and the production supervisor was
Donald F. Schmidt. This book was set in Century Schoolbook. It was composed
by McGraw-Hill's Professional Book Group composition unit.*

*Printed and bound by R. R. Donnelley & Sons Company.*

McGraw-Hill books are available at special quantity discounts to use as premiums
and sales promotions, or for use in corporate training programs. For more information, please write to the Director of Special Sales, McGraw-Hill, Inc., 11 West
19th Street, New York, NY 10011. Or contact your local bookstore.

# Contents

## Part 4    Summaries and Predictions

# Preface

At 4:31 a.m. on January 17, 1994, Southern Californians were suddenly jolted from deep sleep. Soon it was obvious that a powerful earthquake was rumbling and crunching its way across the Los Angeles basin buckling freeways and burying city streets in heaps of dusty rubble. More than 50 people died during the quake, which destroyed $30 billion in personal and public property. Dazed residents numbering in the thousands were left without water, electricity, or telephone service as they huddled in the dark, worried and alone, hoping that rescue forces would soon come.

Telephone lines crashed in sprawling suburbs, but fortunately the earthquake occurred in an era when cellular telephones are everywhere. Cellular phones are based on tetherless technology; they relay voice messages by sending radio transmissions through rugged metal towers sunk into the ground every few miles within densely populated urban areas. Most cellular telephone towers survived the quake to continue carrying cries for help toward police stations, fire departments, paramedics, and other rescue workers. Fortunately, Los Angeles has long been the cellular capital of the western world, with more cellular telephones than any other city, including New York.

When the worst aftershocks finally subsided, local residents learned to their dismay that large sections of their vaunted freeway system had fractured and collapsed. Months, perhaps years, would be required to rebuild that part of the city's infrastructure.

Postquake journeys, it turned out, were taking two or three times longer than they had before. Consequently, many office workers soon began to discover the beauties of electronic telecommuting. Instead of crawling to work on dilapidated freeways, they began to use computer terminals, voice mail, beepers, cellular telephones, and other electronic devices to help them stay at home and do their work.

Even before the Northridge earthquake, cellular telephones had, of course, already begun to alter the American way of life in a hundred dozen different ways. Private businesses, government offices, and ordinary citizens have been rushing to take full advantage of the new tetherless approach. Carry-out meals can be ordered while coming home from work, and business meetings, or heavy dates, can be canceled or changed—on the fly! Stranded motorists report that they feel safer because of cellular phones. And family life seems to work better now that husbands and wives know for sure they can always keep in touch.

Thus we see that the cellular revolution is already sweeping across the American landscape. And now another revolution is waiting for us in outer space. Soon cellular telephone systems will be augmented by huge swarms of satellites whirling around the globe. This book deals with the many interesting facets of that new spaceborne cellular telephone revolution.

Chapter 1 reviews the conventional cellular telephone industry here on earth with emphasis on how it is changing our lifestyles in so many different ways. Chapter 2 examines America's early communication satellites, including active and passive satellites and the ones we have been installing along the geosynchronous arc. Chapters 3 through 5 briefly review today's existing mobile spaceborne communication systems now serving ships, trains, trucks, and planes.

Chapter 6 examines the overall architecture for tomorrow's constellations of mobile communication satellites now being installed or planned for future years. The various tradeoffs necessary to design a viable and efficient multibillion-dollar mobile satellite constellation are also reviewed.

Chapters 7 through 11 examine the various spaceborne mobile communication systems now emerging from the drawing boards. These include Motorola's Iridium constellation and the Russian Globis orbital antenna farm together with Starsys, Orbcomm, and the Globalstar satellite constellations. TRW's medium-altitude Odyssey constellation is also examined together with Ellipso, an elliptical-orbit constellation. Tomorrow's mobile communication satellites will be giving us global cellular telephone services, new avenues for digital data exchange, new paging services, and much, much more. Once they are in place, if another earthquake strikes anywhere in the world, rescue forces will definitely be able to keep in touch.

Chapters 11 and 12 finish the book with capsule summaries and a few imaginative predictions concerning the new electronic technologies likely to take shape in Century 21. Special topics include The Ultimate Personal Computer, tomorrow's multifunctional satellite constellations, and a marvelous new method for eliminating the worrisome swarms of space junk now whirling around the earth.

A major book is invariably a cooperative effort involving combined inputs from many talented individuals. This one, it turns out, was no exception. I would like to thank in particular my darling wife, Cyndy, who suffered through the birthing of the final manuscript with bubbling good humor. In addition, Cyndy, with flashing fingers, typed and corrected the many preliminary drafts up to and including the final version. Her tireless and uncomplaining efforts on my behalf will always be remembered with much affection. So will her wise advice to this easily baffled wordsmith.

I would also like to take this opportunity to thank the artists who used their trusty Macintosh computers to construct the many fine figures that accompany the text: Anthony and Diane Vega, Lloyd Wing, Dick Williams, and Ed Roman.

My agent, Jane Jordan Browne, was, as usual, enormously helpful. Among other things, she carefully modified the contract to help everyone arrive at a mutually beneficial arrangement. And she managed to hold on to the movie rights, too!

Steve Chapman, my editor at McGraw-Hill, was also extraordinarily helpful. During the various stages of production, he worked diligently to keep the project on schedule while still producing a book of professional quality. Finally, I would like to thank the various reference librarians: Nan Paik, Charlotte Baughman, and Alice Hamilton, who helped me locate many key pieces of information to round out the contents of the various chapters. Their efforts to maintain the momemtum of the project are graciously appreciated.

A book is invariably a cooperative effort involving the combined contributions of many talented individuals, and this one was no exception. I sincerely thank all the special people listed here, and in addition, anyone else whose helpful efforts I may have inadvertently overlooked.

*Tom Logsdon*

# Mobile Communication
# Satellites

# The Growing Popularity of Today's Radio-Frequency Communications

# 1

# Expanding Markets for Cellular Telephones

*You could put in this room all the radio telephone apparatus that the country will ever need.*

W. W. DEAN
*Bell Telephone Company President, 1907*

In 1865, when President Abraham Lincoln was assassinated, 12 days went by before the news reached the streets of London. By contrast, in our modern era, whenever anything important happens, people everywhere quickly learn about it as words and pictures ricochet around our planet at the speed of light.

In the late 1980s at a primitive village on the island of Fiji, I received an unexpected lesson concerning today's global distribution of information. The villagers were clothed in colorful native costumes and they lived in little clusters of thatch-roofed huts. So I naturally assumed that I was to be regaled with exotic stories and village lore. What I did not realize was that all of them owned portable radios. So, instead of entertaining me with wondrous tales of native life, they spent the better part of that happy afternoon pumping me for every scrap of information about how I had helped America's astronauts reach the surface of the moon.

As international industries become ever more sophisticated and refined, correct and timely information becomes increasingly important for successful competition and cooperation. Consequently, we have hemstitched the planet with an extraordinarily complicated infrastructure devoted to the efficient distribution of massive quantities of information. Telephone wires carry information. So do microwave relay links and coaxial cables. Information streams toward us from radio and television transmitters and from paging units and cellular telephones. It ricochets through hair-thin optical fibers. And it also comes down toward us from outer space.

America's telephone companies employ more than 650,000 professionals who service and maintain 1.5 billion mi of telephone lines. The value of their plant and equipment totals $264 billion. Each year Americans complete 1.2 billion international telephone calls, many of which are routed through a thin metal daisy chain of commercial communication satellites hovering along the geosynchronous arc 19,300 nautical mi above the earth.

Americans support 9400 radio stations plus 1400 television stations and they purchase 62 million newspapers every single day. Rotary and pushbutton telephones ring incessantly in 94 percent of American households with nearly as many, 92 percent, reporting that they own at least one television set. An even larger number of American households, 99 percent, own at least one radio. We are by any reasonable standard immersed in a sea of information, most of which arrives in our homes in the form of electromagnetic waves.

## Growth Trends for Today's Cellular Telephone Markets

Cellular telephones are the fastest growing segment of America's domestic telecommunications market. Total revenues are relatively small compared with those brought in by conventional telephones. But, for a full decade, cellular telephone sales have been shooting upward at a compound rate of 76 percent per annum. As the graph in Fig. 1.1 indicates, cumulative cellular telephone sales will soon exceed 20 million units with 5 million or so new ones being purchased with every passing year.

The first 10 million domestic cellular telephones were sold within 7 years after practical units first became available. By comparison, it took 9 years to sell 10 million video cassette recorders (VCRs), 22 years to sell the first 10 million fax machines, and 41 years to sell the first 10 million electronic pagers.

"Forecasts of at least 100 million cellular users by the year 2000 are considered conservative," notes industry observer Neil J. Boucher, whose *Cellular Radio Handbook* is being used by armies of neophytes seeking employment and investment opportunities in this hot new field.[1]

## A Short History of Electronic Technology

Carefully modulated electromagnetic waves form the basis of practically every modern method of communication. Electromagnetic waves are waves made up of oscillating, mutually perpendicular magnetic and electric fields that travel through a vacuum at the speed of light—186,000 mi/s. Electromagnetic waves are created when an electric charge oscillates back and forth or is accelerated at a rapid rate.

In 1864 the British physicist James Maxwell predicted the existence of both visible and invisible electromagnetic waves. He also deduced many of their physical properties from a set of partial differential equations he derived. In the late 1880s, Henrich R. Hertz, a German physicist with a highly creative approach, managed to verify Maxwell's elegant predictions by causing an electric charge to oscillate, thus producing electromagnetic waves with much longer wavelengths than visible light.

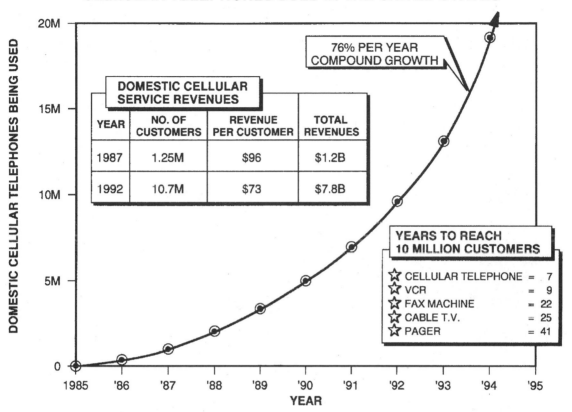

## CELLULAR TELEPHONES SOLD IN THE UNITED STATES

**DOMESTIC CELLULAR SERVICE REVENUES**

| YEAR | NO. OF CUSTOMERS | REVENUE PER CUSTOMER | TOTAL REVENUES |
|------|------------------|----------------------|----------------|
| 1987 | 1.25M | $96 | $1.2B |
| 1992 | 10.7M | $73 | $7.8B |

76% PER YEAR COMPOUND GROWTH

**YEARS TO REACH 10 MILLION CUSTOMERS**

☆ CELLULAR TELEPHONE = 7
☆ VCR = 9
☆ FAX MACHINE = 22
☆ CABLE T.V. = 25
☆ PAGER = 41

**Figure 1.1**  Cellular telephones are the fastest growing segment of America's vast and expanding telecommunications industry with a decade-long compound growth rate averaging 76 percent per year. By 1993 more than 13 million Americans had purchased cellular telephones to be used for an amazing variety of purposes. Some experts are confidently predicting that the global population of cellular telephones will exceed 100 million by the year 2001.

Fifteen years after this landmark demonstration, Guglielmo Marconi, an Italian inventor, combined these two discoveries with a few creative ideas of his own to broadcast the first artificial radio signals through the air. Six years later, in 1901, Marconi's radio equipment was used to send coded telegraph signals across the Atlantic from England to Newfoundland. But high-speed electronic switching devices—vacuum tubes, transistors, modern solid-state circuit chips—were required before successful telecommunication devices could be devised.

The vacuum tube amplifier (Fig. 1.2) was constructed in 1908 by the American inventor Lee De Forest, who inserted a wire-mesh grid between the two electrodes of the Fleming diode developed 3 years earlier by the British inventor J. A. Fleming. A whisper of voltage fed to the grid created a duplicate roar between the two electrodes. Loudspeakers were one quick result. So were practical radio transmitters and receivers.

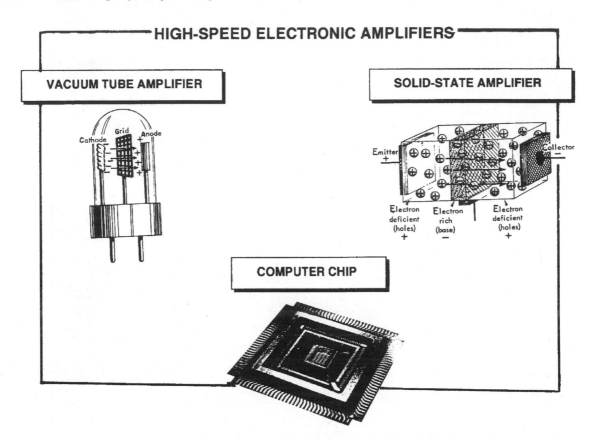

**Figure 1.2**  The vacuum tube amplifier triggered a whole series of technological breakthroughs that made the telecommunications revolution a practical reality. The transistor (solid-state amplifier) is the vacuum tube's kissing cousin in that it also provides circuit switching and signal amplification. The solid-state miracle chip at the bottom of this figure incorporates millions of individual transistors each of which is the functional equivalent of one pickle-sized vacuum tube.

The modern transistor (solid-state amplifier) is a sandwich of semiconductor materials capable of amplifying an electrical current. A schematic diagram highlighting its physical characteristics is sketched in Fig. 1.2. The central portion is manufactured with a surplus of electrons in its crystal lattice structure, whereas the two outer portions are purposely produced with electronic "holes." A *hole* is a place where a negatively charged electron ought to be present in the crystal structure, but is in fact missing.

Like the vacuum-tube amplifier perfected by Lee De Forest, the transistor amplifies weak input signals. A voltage drop is maintained between the two ends of the device so that electrons can be made to migrate across the central portion encouraged or discouraged by relatively weak signals fed into the base. Modern solid-state circuit chips are electronic "layer cakes" of immense complexity. Some versions contain millions of transistors, which are laid down in neat and precise checkerboard patterns.

## A SUDDEN DOWNFALL FOR THE PONY EXPRESS

Advancing technology doomed the Pony Express to inevitable failure, even though Alexander Majors and William Waddle poured nearly $500,000 plus 18 months of meticulous planning and hard work into their highly publicized, but ill-fated, venture. In 1860, Majors and Waddle purchased 500 fast horses and hired 80 young riders to carry the mail at bone crunching speeds over rugged western trails totaling 1966 mi between St. Louis, Missouri, and Sacramento, California. Along the way, the riders stopped at 190 way stations spaced 10 or 15 mi apart.

The Pony Express riders, many with faces still edged with teenaged fuzz, averaged 75 mi/day, mounting a fresh horse at each station. Typically, they were paid $100 to $150 a month to carry ½-ounce letters packed in leather pouches slung over their saddles. At first a half-ounce letter was delivered for $5, but later the rate dropped to only $1 for the same service.

Not without difficulty, horse and rider averaged a rather impressive 10 mi/h along the 2000-mi route with letters arriving—most of the time—within 10 days after they were posted. Unfortunately, even the fastest horses were unable to compete with the electronic telegraph whose pulses whipped across the countryside at 186,000 mi/s.

The Pony Express shut down in October of 1861, 2 days after the transcontinental telegraph lines were successfully strung along a parallel route. During its brief life, the sponsors of the Pony Express received an avalanche of favorable publicity in both domestic and international newspapers, and the young riders were immortalized by Frederick Remington and other western painters. But despite the warm feelings it engendered, the Pony Express did not turn out to be a particularly productive investment. When net receipts were totaled up, its sponsors found that they had lost more than $100,000 in a little over 18 months.

Compared with the primitive vacuum tubes of yesteryear, these devices exhibit some rather amazing capabilities. "If the efficiency and cheapness of the car had improved at the same rate over the last two decades," observes electronics expert Christopher Evans, "a Rolls-Royce would cost about $3, would get 3 million miles to the gallon, and would deliver enough power to drive the Queen Elizabeth II."

## The Electromagnetic Frequency Spectrum

All electromagnetic waves, including radio waves, television waves, radar, visible light waves, x rays, and Gamma rays, are functionally identical and they all travel through a perfect vacuum at precisely the same speed—186,000 mi/s. The various types of electromagnetic waves are distinguished from one another only by frequency and, correspondingly, wavelength. The radio waves broadcast by amplitude modulation (AM) radio stations are about 1000 ft long with a frequency of approximately 1 million oscillations per second (1 MHz). Infrared waves—invisible to the human eye—are 0.001 in. long. They oscillate at $10^{13}$ Hz (10,000 GHz).

A simple bar chart highlighting some of the salient features of the electromagnetic frequency spectrum is depicted in Fig. 1.3. Notice that its logarithmic scales are marked off in powers of 10. Both the wavelengths and the frequencies depicted in this chart vary over 28 orders of magnitude!

## THE ELECTROMAGNETIC FREQUENCY SPECTRUM

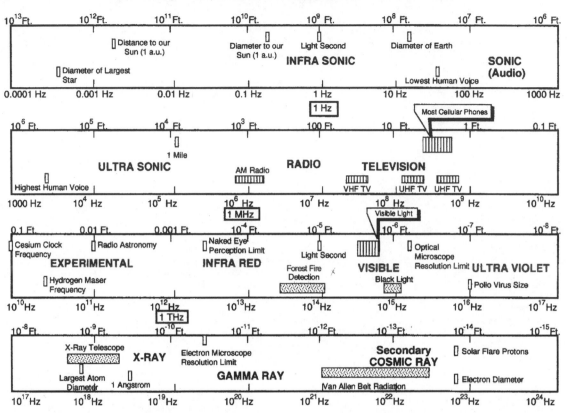

**Figure 1.3**  Visible light, television, radio, radar, x rays, and Gamma rays are all merely different types of electromagnetic waves. They differ only in wavelength and, consequently, oscillation frequency. The electromagnetic frequencies depicted in this figure range from $10^{-4}$ to $10^{24}$ oscillations per second—a total variation amounting to 28 orders of magnitude. Most of the world's cellular telephones operate in the narrow frequency range between 400 and 950 MHz as depicted by the small rectangle sandwiched into the upper right-hand corner of the second band.

For easy reference, various important frequencies are highlighted in the frequency spectrum. The small rectangle on the left-hand side of the third bar, for instance, marks the oscillation frequency of a cesium atomic clock (9,192,631,770 oscillations per second). Cesium clocks are currently used in a voting network to set up the official time standard for the western world. The third bar also exhibits the oscillation frequency of hydrogen masers (1,420,405,751 oscillations per second), which are even more stable and accurate than cesium atomic clocks.

The visible light portion of the electromagnetic frequency spectrum is located on the right-hand side of the third bar. Visible light waves are less than 0.0001 in. long with oscillation frequencies in the $10^{15}$ Hz range (approximately 1 million GHz).

The frequencies most commonly associated with today's cellular telephones are located in the small rectangle on the right-hand side of the second bar. Domestic cellular telephone frequencies vary from about 400 to 950 MHz, although various other frequencies have been used experimentally in America and for routine service in various other countries.

## Conventional Telephone Services

The earliest telephone systems allowed voice communication only between fixed locations such as office buildings, factories, houses, and farms. The majority of early installations were devoted to white-collar office work. Most American homes lacked telephones until the close of World War II when ordinary Americans suddenly seemed to realize that they could hardly exist without being able to communicate with their neighbors.

In the 1920s, telephone company executives scanning the horizon for untapped markets devised some rather wacky and wonderful advertising campaigns. Soon they began to publicize phone-distributed "lullabies to put children to sleep" and "long-distance Christian Science healing." They also offered "sports results, train arrival times, wake-up calls, and night watchman call-ins."

The first big user groups to break free of their telephone tethers were large shipping concerns whose engineers quickly perfected ship-to-shore communication systems, thus allowing well-heeled cruise ship patrons to make stock transactions and act on business decisions while they were at sea. Of course, these so-called radiotelephones were in reality heavy radio transceivers adapted for shipboard use.

## Two-Way Mobile Radio

A conventional two-way mobile radio system does not have the utility and convenience of a cellular telephone. Some of them, in fact, employ simplex voice channels rigged with the familiar "push-to-talk" microphone switches commonly seen in early Hollywood films. But two-way mobile radio is big business, bigger, in fact, than all the domestic cellular telephone systems currently in operation.

As Table 1.1 indicates, conventional two-way mobile radio systems are currently generating an estimated $15 billion in annual revenues per annum compared with $10 billion for cellular telephones and $2 billion for nationwide and local paging services. Of course, mobile services in general are dwarfed by the

TABLE 1.1   Estimated Annual Revenues for Various Sectors in the Telecommunications Business

| Market sector | Estimated annual revenues (billions) |
|---|---|
| Conventional telephone networks | $172 |
| Paging services | $2 |
| Cellular telephones | $10 |
| Conventional two-way mobile radio | $15 |
| Cable television | $14 |
| Home video | $13 |
| Domestic software | $25 |

$172 billion in annual revenues pulled in by our conventional telephone networks—which are a large, mature industry growing at a rather leisurely pace.

A two-way mobile radio system is usually designed by selecting a single channel or a group of channels from a specific frequency allocation for use in a particular geographic region. High-power transmissions are normally used to allow the two-way mobile radio system to cover as much ground as possible. Of course, federal regulations limit the allowable transmission power, so some degree of frequency reuse can be achieved at distant locations.

Two-way mobile radio systems are commonly used in fleet operations: police cars, taxicabs, or delivery vans, for instance. If a vehicle equipped with a mobile radio initiates a call in one zone, its owner must start all over to reestablish contact if the vehicle moves to a new coverage zone served by different transceivers and radio-frequency antennas. Simplex and full-duplex are two popular operating modes for mobile radio systems.

In the simplex mode the same frequency is used for transmission and reception so both parties cannot speak at the same time. Most of today's simplex systems are "voice activated." When one party begins to speak, that direction of transmission has priority until a gap in the conversation is encountered. Unfortunately, voice-operated simplex systems perform erratically and tend to frustrate and confuse both sender and receiver.

The full-duplex mode employs two parallel communications channels to enable both parties to talk simultaneously. Users prefer this mode, but it increases the cost of the system, uses more bandwidth, and requires a bulky and expensive diplexer, which significantly reduces the useful range and coverage areas of the mobile radio communication system.

A quick glance at the electromagnetic frequency spectrum chart in Fig. 1.3 might seem to indicate that choice frequencies should be available in abundance. But, actually, one of the biggest problems facing the global telecommunications industry is the limited availability of usable frequencies. This is true because so many frequencies are already in use and because so many different groups are vying for every smidgen of potentially available frequency spectrum. Often various groups of new users are campaigning to gain access to the same highly desirable frequencies, which cannot be allocated to everyone.

Extensive frequency reuse offers one promising solution. Cellular telephone systems, in particular, reuse the same frequencies numerous times within the same general geographical area. Instead of relying on widely separated high-power transmitters, cellular telephone systems purposely employ large numbers of widely dispersed low-power transmitters with automatic hand-off capabilities to provide seamless communication services to a local group of users.

## The Cellular Telephone Approach

Some cellular telephone users quickly become addicted to the convenience of having instantaneous communication at their fingertips. "It's really like candy," says George Prouty of Essex, Connecticut, "I was making calls from the car while I sat in the driveway!" His daughter Patia, who resides on the Pacific coast in Venice, California, is similarly dumbfounded by the many ways in which she has learned to use her cellular telephone. "It's easier and more convenient than a pager," she notes, "and I feel much safer in the city."

In various countries served by mature cellular telephone systems, 20 percent of all new telephone installations are cellular. In some urban areas, such as Los Angeles and New York, over 500,000 cellular telephone subscribers are paying an average of $80 a month for cellular services.

The earliest proposed cellular telephone architectures date back to the 1940s when a group of engineers at AT&T proposed a high-density mobile system. A cellular system conceptually similar to the one they had proposed was first implemented in Tokyo in 1979. It had a capacity of 4000 subscribers with easy provisions for expansion to 8000.

The heart of any successful cellular system is a computer-controlled cellular switch that can automatically hand-off mobile users from one transmitting base station to another whenever signal quality degrades. The switch automatically assigns a new duplex frequency to such users. Cellular telephone is thus a dynamic communication system that features automatic adjustments and new connections as users move around within the system's coverage region.

## The Marketing Potential of Interlocking Hexagonal Cells

Extensive frequency reuse is possible in a cellular telephone system because low-power transmitters are purposely used to broadcast from relatively unobtrusive base station antennas scattered within densely populated urban areas.

The sketch at the top of Fig. 1.4 features the customary seven-cell configuration commonly seen in marketing brochures distributed by cellular telephone companies to the general public. At the center of each hexagonal cell is a base station consisting primarily of a power source, computer processing devices, and a metal tower with a triangular fixture mounted on its top. Cellular telephone antennas are draped down along the three sides of the triangular fixture. They provide RF coverage for the region surrounding the base station antenna. Each of the seven base stations in the diagram at the top of the figure operates on a different frequency, here denoted by F1, F2, ..., F7. Sequential subscripts denote adjacent frequencies.

Notice how the seven interlocking hexagonal cells fit together in a neat geometrical pattern similar to the one perfected by the lowly honeybee that helped make the 19th century British navy such a formidable fighting force.* This

---

*The two British biologists Karl Vogt and Ernest Haeckel once observed that the British Empire owed its power and wealth largely to bumblebees. They reasoned that the true source of England's influence resided mainly in its superb navy, whose sailors were fed with beef "which came from cattle that subsisted on clover, which would not grow without pollination by bumblebees." (Quote extracted from Tom Logsdon, *The Navstar Global Positioning System,* Van Nostrand Reinhold, 1992.)

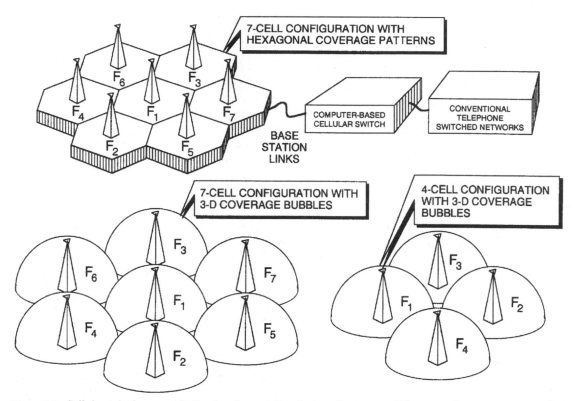

**Figure 1.4** Cellular telephone marketing brochures often feature the seven-cell hexagonal coverage pattern depicted in the top half of this figure. In the real world, however, 3-cell, 4-cell, and 12-cell configurations are nearly as common as the now-familiar 7-cell configuration. Moreover, because of blockages and signal reflections from local terrain features, the coverage patterns associated with real-world cells are, more typically, shaped like amoebas than these precise and regular hexagonal shapes.

rather elegant and precise diagram turns out to be a surprisingly powerful marketing tool. However, the coverage patterns for real-world cellular telephone systems are quite a bit messier and more imprecise.

Seven-cell configurations are used in the industry, but so are 3-cell configurations, 4-cell configurations, 12-cell configurations, and even 20-cell configurations. Moreover, even when a seven-cell configuration is employed, the signals from the individual base stations do not span neat and precise hexagonal cells.

The hemispherical coverage regions sketched at the bottom of Fig. 1.4 are a bit closer to reality, but they are a fiction, too. Neat and clean coverage zones do not exist in the real world because houses, buildings, and natural barriers together with unavoidable sources of RF interference create coverage regions that are shaped more like amoebas than circles or hexagonal cells.

Of course, a large urban area is carpeted with a much larger number of coverage cells than the sample shown in Fig. 1.4. In those cells the same frequencies are reused over and over again. The cells are typically several miles across in regions with moderate population densities with much smaller cells to cover regions with extremely high traffic loads.

Notice how Fig. 1.4 indicates that the various base stations are connected to the computer-based cellular switch, a specifically designed computer telecommunications facility that sets up the proper connections, keeps track of billing charges, and automatically handles any necessary hand-offs. Hand-offs to a new base station are attempted whenever the signal quality degrades as users travel through the cellular telephone coverage area from one cell to another.

Trunk lines connect the cellular switch to the conventional telephone switched network so calls to and from the mobile cellular telephone system can originate from or be directed toward ordinary telephones or cellular telephones located in completely different parts of the country.

## What Is a Cellular Telephone System?

No strict definition of a cellular telephone system is generally accepted by industry professionals, but most experts would agree that it usually entails the following specific characteristics:

1. Extensive frequency reuse with a large number of widely dispersed low-power transmitters located at the base stations

2. Special design features that allow transmitters and receivers to operate in a controlled-interference environment

3. Computer-controlled capabilities to set up automatic hand-offs from base station to base station when the signal-to-noise ratio or transmission distance can be improved to a more acceptable value

Because of their extensive frequency reuse in a small local area, cellular telephone systems can handle a multitude of users. In most urban areas government regulators maintain the proper competitive environment by licensing two separate cellular telephone companies, thus giving customers a choice between competitors should disputes arise.

## Setting Up a Cellular Telephone Call

When a phone call comes into the cellular system from the conventional telephone switched network or from another cellular telephone, the computer-based cellular switch follows the three steps depicted in Fig. 1.5 in setting up the proper connection.

In the first step a call comes in and the cellular switch directs all the base stations under its control to broadcast a paging message directed toward the intended recipient's cellular phone. In the second step the cellular phone being paged acknowledges receipt of the page. In the third step the base station selects a duplex voice channel to handle the call, then signals the particular cellular phone to help set up the desired connection. Similar sequences of operations are initiated to execute a hand-off from one base station cell to another.

## INITIATING A CELLULAR TELEPHONE CALL

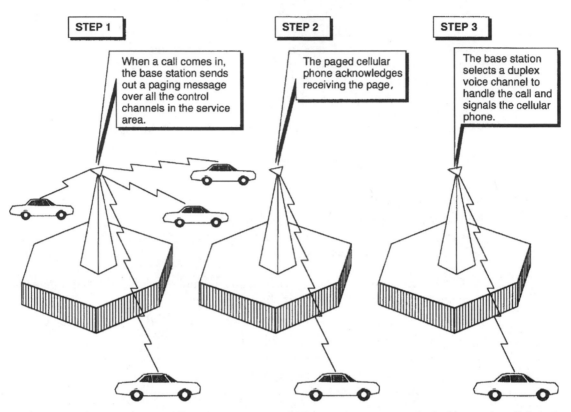

**STEP 1**

When a call comes in, the base station sends out a paging message over all the control channels in the service area.

**STEP 2**

The paged cellular phone acknowledges receiving the page,

**STEP 3**

The base station selects a duplex voice channel to handle the call and signals the cellular phone.

**Figure 1.5** The three major steps followed in setting up a call between a conventional telephone and a cellular telephone are depicted in this figure. In step 1 an appropriate paging message is broadcast to all the control channels in the service area. In step 2 the appropriate cellular telephone acknowledges the page by sending a digital pulse train back to the base station from which the signal came. In step 3 the base station automatically selects and activates a duplex voice channel to handle the call, then signals the appropriate cellular telephone to transmit and receive on the designated frequencies.

Base-station hand-offs figured prominently in the national news on June 17, 1994 when former football star O. J. Simpson fled along the San Diego Freeway in his Ford Bronco to avert arrest by Los Angeles police. His whereabouts were unknown until police officers obtained a court order allowing them to pinpoint the location of his Bronco whenever he made a call on his cellular phone. This was accomplished by interrogating the computer-based cellular switch to ascertain his progress as the vehicle moved from one base station to another.

### Industrial Trends in Cellular Telephone

Throughout the past few years various telecommunication giants have been scrambling to set up strategic alliances in hopes of cashing in on the fastest growing segments of the international telecommunications industry: spaceborne communications, international telephone calls, and cellular telephone. Small wonder they were enchanted with merger possibilities because growth rates in these market segments were expected to shoot through the roof. Revenues for global satellite communication services, for instance, have been projected to increase from $16.4 billion in 1992 to $43.3 billion 10 years later in 2002.

Mobile communications from space and broadcast satellite services also exhibit an exciting potential for runaway growth. Each of them has projected market penetration levels in the $10 billion range. International telephone calls, not including computer data transmissions, amounted to about 20 billion minutes in 1988. By 1993 that figure had grown to 41 billion minutes and by 1995 it is expected to reach 60 billion minutes.

One of the most exciting mergers occurred in August of 1993 when AT&T purchased McCaw Cellular for an estimated $12.6 billion in AT&T stock, thus making a bold and costly move into mobile communications. At the time McCaw Cellular, with its 2.5 million subscribers, was the largest cellular operator in the United States.

Industry observers noted that the match-up allowed AT&T to dive into the cellular industry which, that year, racked up revenues of $7.8 billion, up 38 percent from the previous year. The deal also allowed AT&T to move one step closer to its corporate goal of allowing people everywhere to communicate "easier and faster by phone, computer, hand-held devices, and other sophisticated technologies," as one industry observer pointedly observed.

The baby Bell companies immediately objected to the buyout, claiming that the new strategic alliance created unfair competition because it allowed AT&T to handle local calls, thus eliminating the usual access fees that were providing about 25 percent of their annual revenues. Other cellular companies were worried because the multibillion dollar deal put them in direct competition with an industry giant intent on carving out a larger segment of the evolving telecommunications market. Loss of a few prime customers could be a big blow to AT&T's new competitors. Pacific Bell, for instance, pulls in 50 percent of its residential toll revenue from just 10 percent of its high-volume telephone customers.

## Moving Toward Global Cellular Standards

The world's most popular cellular telephone systems are AMPS (Advanced Mobile Telephone System), developed in the United States, and TACS (Total Access Cellular System), developed to serve various European countries. AMPS currently has about 62 percent of the users in the world, whereas TACS is a distant second with 13 percent.

The American AMPS is an 800-MHz system with 30-kHz channel separations. Each cell handles 832 frequency modulation (FM) channels with digital frequency shift keying for the control-channel modulations. AMPS is presently being used in 37 different countries.

The TACS system operates in 21 countries at 900 MHz with 920 channels separated by 25 kHz. Like the AMPS system, TACS uses FM analog voice-channel modulations with digital frequency shift keying for the control channels. Both analog and digital modulation techniques are described in detail later in this chapter.

Third World and emerging countries are also benefiting from the new cellular technologies. Costa Rica, Ghana, Mexico, Pakistan, Paraguay, and the Philippines all have cellular systems. In fact, their average monthly bills are higher than the bills of users in more highly developed countries where other cheaper means of communication are readily available. One 1991 study showed that the average monthly bill for Third World subscribers was $115 compared with $85 for their counterparts in the United States.[2]

Third World countries have a strong pent-up demand for telecommunications services, and cellular is a definite candidate for filling their needs. As cellular expert Alf Humphries put it: "Cellular phones give poor countries a quick, relatively cheap way to obtain late 20th century communications technology."

At present, per-minute rates for cellular telephones are higher than they are for conventional telephone services, but that situation may not persist indefinitely. One intriguing study revealed that the installed base of equipment is not any higher for cellular. However, conventional phone companies have been in existence for many years, so they have had longer to pay for larger portions of their telecommunications equipment.

## Cellular Telephone System Design

Designers of cellular systems must make a number of engineering choices centering around such practical matters as frequency choices, signal modulation techniques, site selections, antenna characteristics, and the like. In this section we will discuss some of these major engineering decisions and the various factors that tend to nudge them in one direction or the other.

## Frequency Selections

Most of today's cellular telephone systems operate in a rather narrow frequency band ranging from about 400 to 950 MHz. The corresponding wavelengths are roughly 1.0 to 2.5 ft long (with the lower frequencies being associated with the longer wavelengths).

Officials at the Federal Communications Commission (FCC) were driven to the 800-MHz frequency allocation for cellular telephones by quite practical considerations. Most of the frequencies below 800 MHz were already heavily occupied by other users such as FM radio (at about 100 MHz) and air-to-ground transmissions (118 to 136 MHz). Consequently, only the skimpiest scraps of the frequency spectrum were available below 400 MHz.

Higher frequencies—those in the 10-GHz band and beyond, for instance—were not a very practical choice because falling raindrops and oxygen absorption create severe propagation losses, signal distortion, and signal fading. *Multipath* occurs when electromagnetic signals are reflected from nearby objects, such as buildings or trees, thus smearing the primary signal coming along a more direct route from transmitters to receiver. The annoying "ghosts" that sometimes appear on your television screen are caused by multipath reflections as the broadcast signals bounce off your neighbor's house.

The 800-MHz portion of the frequency spectrum had originally been allocated to educational television. But, when so many homes were wired for cable, this channel space was, to some extent, reallocated, thus opening up the 800-MHz band for America's mobile telephone services. However, as Fig. 1.6 indicates, some of the cellular telephone systems in other parts of the world are being implemented at slightly different frequencies. Canada's Aurora 400 system operates at 400 MHz, for instance, and the Scandinavian countries operate at 450 and 800 MHz.

Various other important frequencies are also marked for easy reference in Fig. 1.6. These include the 9.192,631,770-GHz oscillation frequency of cesium atomic clocks and the 1.402,405,751-GHz frequency of hydrogen masers. The figure also highlights the 150- and 400-MHz frequencies used by the Navy's low-altitude spaceborne Transit navigation system and the two L-band frequencies of the Navstar Global Positioning System. The 30-GHz rain attenuation and the 60-GHz oxygen attenuation frequencies are also pinpointed on the spectrum chart.

Cellular telephones are purposely designed to operate in an environment filled with potential sources of interference, while still producing satisfactory results. Interferences and blockages come from a variety of sources including other base stations in the system, stray radio waves, foliage attenuation, and blockages from structures, hills, and mountains. Proper design and placement of the antennas used in a cellular telephone system can minimize these worrisome effects to some extent, but blockages and interference can never be entirely eliminated.

## Signal Modulation

Most of today's cellular systems employ analog modulation techniques in which their electromagnetic carrier waves are modified by a direct, smoothly varying analogy of the human voice being transmitted to the intended recipient. AM (amplitude modulation) and FM (frequency modulation) radio stations employ similar analog modulations. In fact, AM and FM are the two most common types of analog modulations. Frequency modulation is the preferred approach for transmitting human voices to cellular telephones, whereas digital modulations are used for the cellular system's control channels that send paging messages out to the users and handle other data exchange functions.

Figure 1.7 highlights some of the more important characteristics of the AM and the FM modulation techniques. For both approaches, a high-frequency sinusoidal carrier wave is modulated by a voltage curve produced by sending the announcer's voice through a studio microphone.

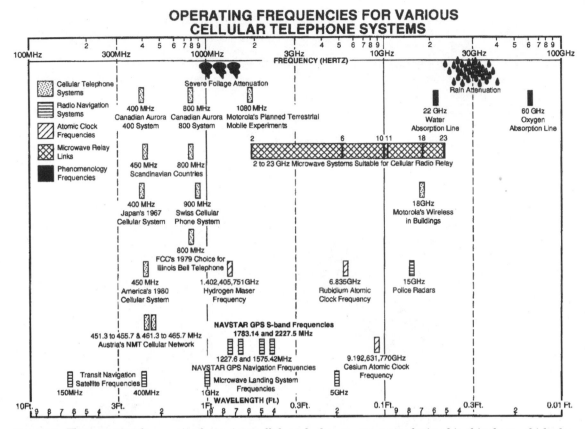

**Figure 1.6** The operating frequencies for various cellular telephone systems are depicted in this chart, which also highlights the frequencies of various other devices such as radionavigation transmitters and atomic clocks. Foliage attenuation hampers the operation of cellular telephones, most of which broadcast their modulated signals somewhere in the range between 400 and 950 MHz. Attenuation due to raindrops and oxygen molecules occurs primarily at frequencies of 10 GHz and beyond. Some cellular telephone systems link their base stations to the cellular switch using microwave relay links in the 2- to 23-GHz portion of the frequency spectrum, where raindrops can, on occasion, create attenuation problems.

For AM broadcasts, the voltage curve creates local *amplitude* variations in the curve actually being broadcast. For FM broadcasts, the voltage curve creates local *frequency* variations in the curve. Regions of high amplitude in the voltage curve produce subtle *increases* in the broadcast frequency. Regions of low amplitude in the voltage curve produce subtle *decreases* in that same broadcast frequency. Any AM or FM radio that picks up the appropriate modulated curve demodulates it to restore the radio announcer's voice more or less to the way it sounded in the studio.

Cellular telephone systems employ FM techniques for their voice transmissions because, in the noisy, electromagnetic environment in which they operate, FM modulations tend to produce more static-free reception for the receiver.

# COMMON ANALOG MODULATION TECHNIQUES

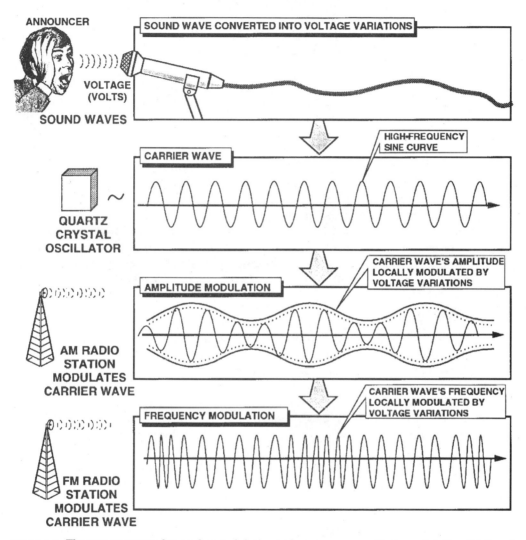

**Figure 1.7** The two most popular analog modulation techniques are amplitude modulation (AM) and frequency modulation (FM). Sounds coming into the announcer's microphone at an AM radio station create continuous voltage variations, which are then superimposed on the station's regularly spaced sinusoidal carrier waves. The AM modulation technique creates localized *amplitude* variations in the carrier waves actually broadcast to the station's listeners. For FM broadcasts, the voltage variations coming from the microphone create localized variations in the *frequency* of the carrier waves.

## Site Selection and Cell-Splitting Techniques

Cellular telephone engineers work long and hard to position their antennas properly in hopes of providing the most interference-free services over their entire coverage regime. Most cellular antennas are positioned on specially built towers or they are attached to the sides of buildings. Some are designed to be highly directional; others radiate their signals 360 degrees toward all points of the compass.

The various base stations in a cellular system always create quite a bit of mutual interference. This unavoidable effect can be mitigated to some extent by careful site selection, choosing the proper heights for the various transmission towers, and by angling the antenna elements upward or downward to modify their coverage zones.

When the calling traffic within a particular zone increases, the cell can be split into smaller cells so the transmission frequencies in that sector can be reused much more often. When a big cell is split into smaller cells, several new base stations and antennas must be installed. Figure 1.8 shows how an old, overloaded cell can be split into seven new cells. In actual practice an old cell is more commonly split into three or four new cells instead. In this case, each of the new cells has about half the diameter of the old cell so four of them, properly distributed, can handle approximately the same coverage zone.

## Hand-offs to a New Base Station

Each base station constantly monitors the signal strength and/or the signal-to-noise ratio of the modulated signals it is receiving from each mobile user. When the quality of this signal drops below certain preselected norms, the base station sends a request to the centralized cellular switch asking it to attempt a hand-off. The switch then automatically asks each adjoining base station to scan the frequency being used, then report the quality of the signal it is picking up. If substantial improvements can be made, the switch automatically orders a hand-off.

Most lost calls occur during hand-offs, so the design engineers attempt to minimize the number of hand-offs that occur. However, this is not always possible because many of the users are mobile, so they are constantly driving outside the range of one base station and into the vicinity of another. Pedestrians can also move around enough on foot to encounter signal blockages. Theoretical calculations and field-test experiments can help system designers determine how often hand-offs are likely to occur.

The results of one such study are summarized in Fig. 1.9, in which the call length is plotted against the hand-off probability. For this particular case, a user who engages in a 9-minute mobile conversation has a 50 percent probability of encountering at least one hand-off before the call is terminated. Fortunately, most cellular conversations are shorter than that. In Southern California, for instance, the average mobile conversation lasts less than 3 minutes, and the average conversation from a hand-held portable cellular telephone spans only about 85 seconds.

## SPLITTING A CELL TO SERVE A HEAVIER TRAFFIC LOAD

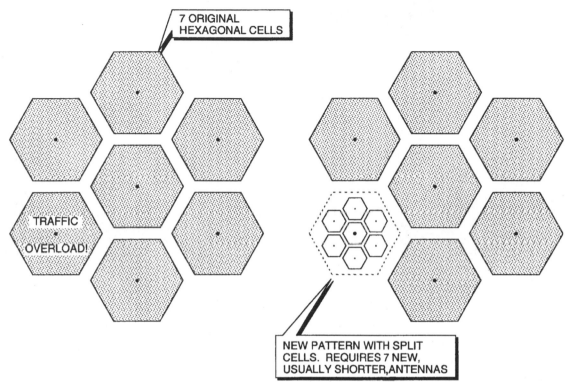

7 ORIGINAL
HEXAGONAL CELLS

TRAFFIC
OVERLOAD!

NEW PATTERN WITH SPLIT
CELLS.  REQUIRES 7 NEW,
USUALLY SHORTER, ANTENNAS

**Figure 1.8**  As the number of subscribers increases, some of the cells in a cellular telephone system will gradually become overloaded with calls, thus reducing signal quality and increasing the probability of service interruptions and lost calls. When this happens, "cell splitting" can be used to reduce service demands on any overloaded cells. A cell is split by subdividing its coverage region into a set of smaller cells each served by its own antenna-equipped base station. To avoid interference with other nearby cells, the smaller new cells are usually served by antennas mounted on shorter towers and operated at lower levels of power.

### Data Link Selection

A cellular telephone system includes a group of widely dispersed base stations connected together through a cellular switch. Microwave relay links often form the necessary connection between the base stations and the remote cellular switch, but other approaches are also possible including optical fibers, leased wirelines, RF transmissions, and even orbiting satellites. Most microwave systems suitable for interlinking cellular stations operate in the frequency band ranging from about 2 to 23 GHz.

### Modern Cellular Telephone Equipment

Cellular telephones come in a variety of different configurations. The best choice for a heavy driver may be the mobile telephone or *car phone* permanently mounted in the vehicle. A mobile telephone can deliver a powerful sig-

## THE PROBABILITY OF A CELLULAR PHONE CALL REQUIRING AT LEAST ONE HANDOFF

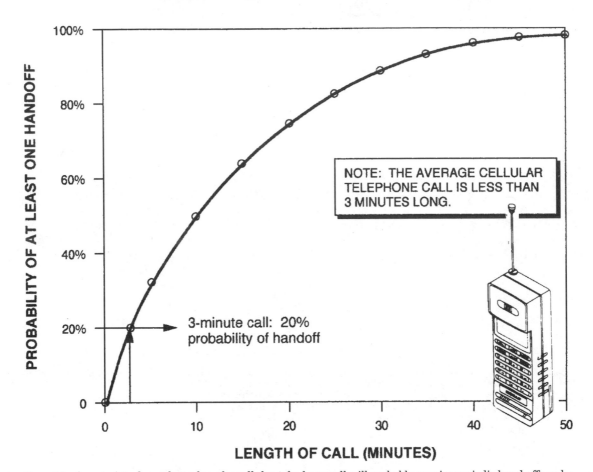

**Figure 1.9** A motorist who makes a lengthy cellular telephone call will probably require periodic hand-offs as he or she travels from the vicinity of one transmitter to another. Poor-quality reception and lost connections tend to occur during these critical hand-off intervals. As this semiempirical graph indicates, approximately half of all 9-minute conversations will require at least one hand-off before the call is completed. Fortunately, most cellular telephone conversations are less than 3 minutes long, so hand-offs occur only about 20 percent of the time.

nal for a long time because it draws its power from the car's electrical system. Most mobile units put out a full 3 watts of RF output power. Some luxury automobiles offer models with hands-free operation rigged so the sound comes through the car's stereo speakers. Some have limited voice recognition capabilities. Users can issue simple commands such as: "call wife" or " hang up" and the mobile phone will respond appropriately.

Those who need to take their cellular phones from car to car or from car to construction site may opt for a *transportable phone* consisting of a handset attached to a rather heavy and bulky transceiver and battery pack. The transportable cellular telephone, which weighs around 5 lb, can draw power either from a car's cigarette lighter or from its own batteries. Transportable models are packaged in two ways. The hardback transportable (Fig. 1.10) and the soft pack or "bag phone" (Fig. 1.11).

The smallest model, a hand-held portable phone (Fig. 1.12), can be carried in pocket or purse. Hand-held portables combine transceiver, antenna, and battery in a single package that resembles a television remote control unit. The biggest ones weigh around 1.5 lb, but some models tip the scales at less than 0.5 lb. The smallest models tend to be the most expensive.

Portable units are convenient, but their power levels are often as low as 0.6 watts, their signal quality and coverage areas sometimes leave much to be desired, and their talk time is strictly limited, typically 2 hours or less. Frequent battery charges are necessary for the portable units, some of which require up to 15 hours to pick up a full charge.

Modern cellular phones include a number of convenient features and add-on options. These include signal-strength meters, battery charge-level meters, one-touch dialing for frequently used numbers, fax-compatible connections, paging capabilities, and the like.

## Marketplace Responses

The typical cellular telephone user is an upper-income professional working for a small- to medium-size company, typically employing 100 workers or less. A 1993 marketing survey conducted by Cellular Marketing, Inc. indicated that 34 percent of domestic cellular users earn more than $100,000 per year, with another 33 percent bringing home between $50,000 and $100,000. About 85 percent of cellular's clients are men with an average age of 40 years. The three most popular urban areas for cellular telephones are Los Angeles, New York, and Chicago, in that order.[3]

In 1991 more than 26,000 full-time employees made their living in America's cellular telephone industry, which at that time served 7.5 million customers with low-power signals radiated from 7800 base stations.

## Customer Concerns

Most cellular clients seem to be reasonably satisfied with the services they receive, but they do have a few nagging complaints. Their most common complaints center around hand-held battery life coverage and call charges. Hand-held battery life typically varies from 30 to 90 minutes. When a battery is fully drained, some company-furnished chargers require up to 15 hours to build it back up again. Consequently, many dedicated portable phone users purchase two batteries and interchange them on alternate days.

Coverage areas for hand-held units can be quite limited, a problem that becomes more severe when the user is in an elevator shaft or on the ground floor of a high-rise building.

**Figure 1.10** A hardback transportable cellular telephone can be operated in a car or truck—or it can be removed for portable operation under its own battery power. When a transportable unit is plugged into the vehicle's power source, it is operated at its maximum power level of 3 watts. In the portable mode, its RF power level is usually adjustable. This allows knowledgeable users to make real-time tradeoffs between signal quality, coverage area, and battery life. (*Courtesy of OKI Telecom.*)

## A TYPICAL SOFT-PACK CELLULAR TELEPHONE

**Figure 1.11** A soft-pack cellular telephone, which is also referred to as a "bag phone," is essentially a transportable unit with batteries and other equipment modules packaged in an attractive and convenient shoulder bag. This particular model provides 45 minutes of continuous talk time at its maximum level of 3 watts continuous RF output power. Optional extras include hands-free operation and an electronic digital data interface to allow for the two-way exchange of computer data. (*Courtesy of Motorola.*)

The monthly bills for cellular telephones are paid by the user's parent company in 78 percent of the cases, but mobile cellular phones are, nevertheless, a price-sensitive product. When purchase prices drop, large numbers of new customers come online. However, when the first monthly bill arrives, this initial spurt of enthusiasm may suddenly damp out. Such buyers frequently cancel at that point or even default on their bills. In the past few years, America's cellular phone prices have dropped from a few thousand dollars to only about $300, where they seem to have temporarily stabilized.

As a cellular system grows larger and larger, service quality usually declines. This is true because as the system approaches its saturation point, call drop-

**Figure 1.12** Hand-held portable cellular telephones weighing 8 oz or less are currently available in today's commercial marketplace. This compact model weighs 7.7 oz and occupies a volume of less than 12 in³. During use, the mouthpiece flips down to uncover the keypad control buttons. Thus, it operates somewhat like the personal communicators used by the officers and crew aboard Captain Kirk's starship *Enterprise*. (*Courtesy of Motorola.*)

outs, message blocking, cross talk, and hand-off failures all tend to increase due to mutual interference between the various base stations. Fortunately cellular customers are gradually learning to appreciate the fact that cellular is a radio system intrinsically different from conventional telephone. As such, its users encounter definite tradeoffs between mobility and service quality.

In some countries, such as Hong Kong and Singapore, 80 percent of all cellular telephones are portable hand-held units. But in the United States only about 19 percent are portables. Another 29 percent are transportable units, but the largest fraction, 52 percent, are mobile phones permanently mounted in cars and trucks.

However, industry sources have recently been reporting substantial growth in the sales of portable and transportable units. This growth trend will probably result in increased customer complaints because smaller, lower-power units tend to provide reduced coverage, lower signal quality, and shorter power-plant life.

## Safety Considerations

Many users purchase cellular phones so they will feel safer, especially if their jobs take them into crime-ridden urban areas. Ironically, the cellular telephones that are supposed to be keeping them safer may be, themselves, a hazard, according to some vocal critics. In 1993, the husband of a cancer victim charged that cellular telephone usage had caused the brain tumor that killed his wife.

The Cellular Telecommunications Industry Association has pledged to spend $15 to $25 million to study the issue anew. But, in the meantime, they strongly maintain that research to date has shown overwhelmingly that the radio transmissions from cellular telephones pose no health risk.

Cellular conversations are carried by radio waves similar to those used for commercial television broadcasts, but with thousands of times less power. Their RF power levels are, in fact, similar to the power level emitted by a small flashlight.

The National Council on Radiation Protection and Measurements has set safety limits for localized occupational exposure at a level of 1.6 W/kg for 30 minutes' duration or more. A portable cellular telephone produces only 0.45 W/kg over the same duration. This is only about one third the level experienced by users of citizens band radios and one seventh the level experienced by the users of police radios.

Sid Deutsch, visiting professor of electrical engineering at the University of South Florida, points out that 17,500 Americans get brain cancer each year and that 1 in 50 Americans is a cellular telephone user. Thus, on average, we can expect that about 350 American cellular telephone users will develop brain cancer in any given year. Consequently, regardless of the findings of the Cellular Telecommunications Industry Association, this controversy will likely simmer for a long time to come.

## Data Services

"Eventually 90 percent of the people who are working will be tetherless," says Ernest von Simpson, Senior Partner of the Research Board of New York City.[4] Moreover, in his opinion, a large percentage of the tetherless people will not be exchanging voice messages; they will be linked together by digital data.

Today data transmission over ordinary telephone lines is growing at 30 percent per year. And, on certain specific routes, data, not voice, constitutes the bulk of traffic. Between the United States and Japan, for instance, fax messages are exchanged on as many as 90 percent of all telephone connections. Combine data with tetherless transmissions and you have the makings of a telecommunications revolution.

"Soon to come: tiny radio modems no bigger than a credit card that will slip into laptops and smaller devices and link them to wireless networks," says *Fortune* magazine reporter Andrew Kupfer. Salesmen, construction workers, meter readers, police officers, and a broad spectrum of off-site workers will be linked to their offices by wireless data communications. One official from BIS Strategic Decisions of Norwell, Massachusetts, believes that, "By the end of the decade, wireless data revenues in the U.S. will hit $10 billion a year."[5]

Of course, many cellular telephone customers already receive limited amounts of digital data in the form of paging. About a third of cellular customers are, in fact, also customers for paging services. This dual-service trend will likely continue because the FCC recently decided to allow paging over cellular telephone systems as a standard feature.

A simple paging device, today, costs about $100, but more sophisticated units with alphanumeric messaging capabilities can go for $500 or more. Centralized voice mail services are also being provided by some innovative cellular systems. Cellular companies that have added this feature find that about 30 percent of their customers are willing to pay the extra fee.

### Acoustic Couplers

Data services are gradually becoming a standard way of using cellular telephones. Unfortunately, a cellular system operates in a rather harsh electromagnetic environment awash with interference, so disruption of data commonly occurs. As with ordinary telephones, digital data is sent over cellular telephones using a modem.

A modem is an electronic device that converts the binary 1s and 0s that come out of a digital device (laptop computer, fax machine) into a corresponding analog signal. At the receiving end, a second modem is used to make the reverse conversion. The simplest modems (Fig. 1.13) employ frequency shift keying to generate two tones. One of the tones represents a binary 0; the other represents a binary 1.

### Cellular Digital Packet Data

Craig McGaw, dynamic mastermind of McGaw Cellular, Inc., has a dream for the future of digital data in which voice and data will be streaming over cellu-

## ACOUSTIC COUPLERS

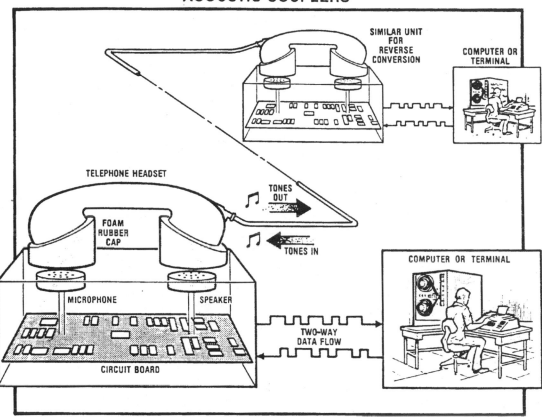

**Figure 1.13** Acoustic couplers allow ordinary telephones and today's analog cellular telephones to transmit digital data to and from distant computers. A second modem at the other end reverses the process to restore the original computer pulses. Advancing technology will allow similar modems to be built directly into the next generation of cellular phones.

lar links to mobile receivers who will be capable of displaying text as well as voice. His concept is called Cellular Digital Packet Data (CDPD).

Today's cellular telephones employ analog modulations which are highly susceptible to interference. Their transmissions provide acceptable quality for voice, but they tend to be ill-suited to the transmission of data in quantity. That is why Craig McGaw slices his data up into small packets with "handshake" acknowledgments at the other end at the time each packet is successfully received. No hand-offs are allowed during the transmission of a packet. If, for any reason, a packet fails to get through to the other end, the CDPD system merely sends it again.

CDPD is essentially a separate transmission device co-located with the conventional voice hardware at the cell sites in a cellular telephone system. So new equipment will be necessary to handle the data services. Industry experts estimate that modifying the 2225 cell sites in McGaw's domestic network will cost in excess of $250 million.

---

### RHYTHMIC TELEGRAPH SIGNALS CLICKED OUT BY SAMUEL MORSE

On Friday, May 24, 1844, an ex-portrait painter named Samuel Morse sat in a straight-backed chair in the sweltering heat of the U.S. Supreme Court Chamber in Washington, D.C., calmly clicking out staccato rhythms on a strange metal contraption. A single strand of insulated wire connected his mechanism with a similar device in Baltimore's Mont Clair railway station 41 mi away. There a knot of tense technicians huddled around a mechanically actuated pencil stub, which flawlessly duplicated Morse's coded message.

To most of the curious onlookers that entire demonstration must have seemed like an extraordinarily complicated technique for painting wavering lines on yellowed sheets of paper. But to the technicians who knew the code, the "dots" and "dashes" making up the message were relatively easy to decipher:

```
· _ _   · · · ·   · _   _   · · · ·   · _   _   · · · ·   _ _ ·   _ _ _   _ · ·
  W        H      A    T      H      A    T      H       G       O       D

· _ _   · _ ·   _ _ _   · · _   _ _ ·   · · · ·   _   · · _ ·
  W       R       O       U       G       H      T      ?
```

Samuel Morse had never been a noteworthy portrait painter, but because he had planned his demonstration with such a dramatic flair, investors with big money and powerful influence were suddenly interested in his clattering mechanism. Within a few short years, similar strands of telegraph wire were carrying coded messages to thousands of destinations all across the country. And two decades later, when swarms of Minie balls began whizzing through the air at Vicksburgh and Harper's Ferry, President Abraham Lincoln was able to receive regular telegraph dispatches pinpointing the progress of America's brutal and bloody Civil War.

---

## Digital Implementations for Tomorrow's Cellular Telephones

We are surrounded by digital devices—digital wristwatches, compact discs, digital supermarket scanners, digital odometers, digital bathroom scales, not to mention digital computers. And yet, today's cellular telephones use analog modulations similar to the modulations used by FM radio stations, which have been in existence for several decades. Not to worry. Digital cellular telephones are definitely on the way!

The primary motivating factor pushing today's designers toward digital implementations for cellular is that, in many urban areas, analog cellular is approaching the saturation point. Many of today's experts are also convinced that, with digital, signal quality will improve.

In a digital modulation technique, the analog voltage curve representing the human voice is broken up into thin vertical stripes with each amplitude being represented by a string of binary digits (1s and 0s). The binary digits are then broadcast to the intended recipient where the process is reversed to reconstruct the original voice tones. Figure 1.14 shows how the digitizing process is accomplished.

## DIGITAL MODULATION TECHNIQUES

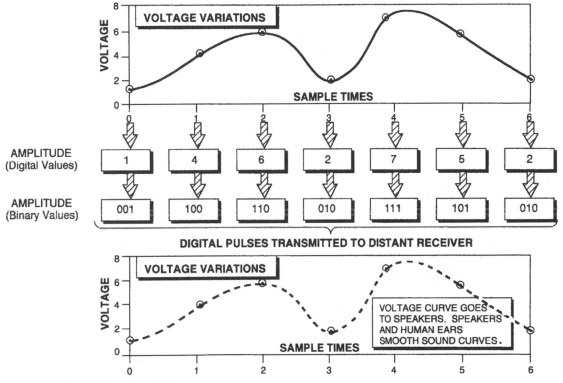

**Figure 1.14**   In this digital modulation system, the original analog signal, such as a human voice, is converted into a series of binary pulses that are then transmitted to a distant receiver. In this simple implementation, each specific amplitude in the voltage curve is represented by a three-digit binary number ranging in value from decimal 0 to decimal 7. When the distant receiver picks up the binary pulses, it uses them to reconstruct an analog version of the original signal being sent.

In the telecommunications industry various methods for transmitting the binary pulse trains have been used, but for future generations of cellular phones, two principal methods are on the verge of being implemented:

1.  Time division multiple access (TDMA)
2.  Code division multiple access (CDMA)

In the *time division multiple access* approach being implemented in conjunction with cellular, the calls are chopped into time slices 1/150th of a second long and intermixed with other similar cellular transmissions. Industry experts estimate that as many as six digital TDMA calls can be interleaved to travel over a single cellular channel. This is possible, in part, because of the momentary snippets of silence in most telephone conversations.

In the *code division multiple access* approach, every call is spread over the entire band of cellular frequencies. The calls are encoded using prearranged binary pulse sequences at both ends of the transmission, so only the intended recipient can separate the voice message from the inevitable background noise. Initial versions are expected to have 10 to 15 times the call-handling capacity of today's analog cellular phones over the same bandwidth.

## The Time Division Multiple Access Approach

The first-generation digital TDMA cellular systems, now being implemented in selected regions such as Los Angeles and Atlanta, are roughly five times more complicated than the analog cellular systems now in use. They will require a great deal of computer memory and computer processing capability, but they do conserve spectrum bandwidth.

The initial digital cellular phones are retailing for $700 to $1500 each, except where they are subsidized by cellular telephone companies. Subjective voice quality will be about the same as today's analog systems, but battery life is greatly improved. Digital cellular will also enhance user privacy and improved digital data-handling capabilities. Later units will provide enhanced services such as voice-recognition and voice-messaging capabilities.

## The Code Division Multiple Access Approach

The CDMA approach will follow the TDMA implementations because the necessary cellular telephone technology has not yet been fully perfected. Military aerospace engineers pioneered the use of CDMA messaging systems, primarily because CDMA provides improved data security and privacy for the intended military users. But those same engineers are convinced that improved spectrum efficiencies can be achieved by CDMA implementations in the cellular world.

Figure 1.15 shows how the CDMA technique is implemented in a cellular telephone system. Both transmitter and receiver are rigged to generate the same prearranged binary pulse train running across the top of the figure. This pseudorandom sequence of binary +1s and −1s is multiplied bit-by-bit by the binary +1s and −1s present in the voice stream. The product sequence (bottom of the figure) is then broadcast for reception by the intended cellular recipient who also generates the prearranged sequence of binary +1s and −1s running across the top of the figure. By comparing the sequence of binary digits it actually receives with the sequence it is currently generating, the cellular telephone can "decode" the voice bits that were multiplied by the original pseudorandom binary pulse train.

Every single active cellular telephone in a CDMA cellular system transmits its voice messages on exactly the same frequencies as its counterparts. So how can the various messages be distinguished one from another? Each transmitter uses a different prearranged pseudorandom pulse train when it modulates the signals being transmitted. In other words, the various transmissions are distinguished by their pseudorandom codes.

## MODULATION TECHNIQUES FOR CODE-DIVISION MULTIPLE ACCESS

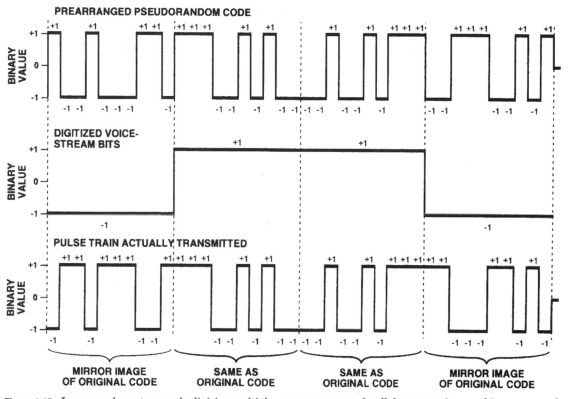

**Figure 1.15** In a spread-spectrum code-division multiple access system, each cellular transmitter and its corresponding receiver are programmed to generate the same prearranged sequence of binary pulses here represented as +1s and −1s. The bits in the prearranged sequence are multiplied by the corresponding +1s and −1s in the digitized voice stream—which has a much lower chipping rate. When the cellular telephone receives the product sequence, it restores the original voice-stream bits by comparing the bits it actually received with the prearranged binary pulse sequence it is currently generating.

Every transmitted message is spread over the entire allowable bandwidth. The result is a so-called *spread-spectrum* signal. It has been given this name because it occupies much more bandwidth than necessary in terms of its raw message content.

The pseudorandom pulse trains used in CDMA spread-spectrum systems are carefully chosen so they are mutually orthogonal. Because of this special property of orthogonality, the cellular telephones in the system are extremely unlikely to lock onto the wrong conversation even for a fleeting instant.

In effect, the cellular phone in a CDMA system "anticipates" what the next voice pulse will be (either a binary +1 or a binary −1). If it is wrong, it will encounter an adjacent sequence of pulses that have been inverted (+1s where −1s ought to be, and vice versa). The cellular phone is thus, to some degree, like a paying customer at a classical music concert trying to pick out the sounds being made by a particular instrument. By anticipating the melody, the listener is able to do it successfully even though the other instruments in the orchestra tend to mask the anticipated sounds.

Spread-spectrum CDMA modulations are used by military communication devices, commercial communication satellites, and in today's most advanced space-borne navigation constellations such as the American Navstar and the Soviet Glonass. Some of the intrinsic advantages of a spread-spectrum implementation can be listed as follows:

1. Messages can be conveniently directed to specific recipients among multiple users.

2. High-resolution ranging is possible for position-fixing and navigation systems.

3. Easy rejection of natural background noise and broadcast jamming signals can be accomplished.

4. Signal hiding is possible for privacy and data security.

For tomorrow's cellular telephone systems all four of these advantageous characteristics may ultimately turn out to be of practical benefit.

## Digital Cellular Systems Currently Being Implemented

Early in 1994, PacTel Cellular (now Airfone) unveiled plans for installing new digital cellular systems in four cities: Los Angeles, San Diego, Sacramento, and Atlanta. The first few installations, which will rely on Motorola equipment, are coming into Los Angeles at a cost of $70 million.

The Airfone System is incompatible with L.A. Cellular's digital approach introduced in late 1993. Some observers are afraid that the two systems will seriously interfere with one another as they grow and expand because neither of them is sufficiently mature for successful introduction. "Everyone is so obsessed with rushing the technology to market that the interference issues won't have been resolved," says Herschel Shosteck, a cellular specialist from Silver Spring, Maryland. "Pushing it prematurely is like trying to make a major league hero out of an 11-year-old sandlot player."

However, the executives from Airfone and L.A. Cellular, who are putting up the money, seem entirely satisfied with the safety and security of their investment. "The customer ultimately will have to decide," observes Susan Rosenburg, a spokesperson for Airfone's marketing division.

## Emerging Concepts for Tomorrow's Personal Communications

Cellular telephone systems have been experiencing a solid decade of extraordinary marketplace growth. Consumer markets have, in fact, grown so rapidly that cellular phone companies have difficulty recruiting and training enough qualified personnel. Expansion will likely continue at a rapid pace, but competing technologies are also poised to capture their own market share.

Personal communication networks serving pedestrians in crowded locations will likely be online within a few short years. They will feature very low-power and inexpensive shirt-pocket telephones serving shopping malls, airline terminals, office corridors, and crowded city streets. The architecture of the personal communication networks will resemble the one used by cellular telephones—separate "cells" each served by its own transmitter. But both phones and transmitters will be ultrasmall and the base stations will be spaced much closer together. Many analysts have suggested a 60-GHz transmission frequency because it is readily available and because its limited transmission range will permit extensive frequency reuse. Other experts, including those at Motorola, are eyeing the 1850- to 1990-MHz band, which they plan to modulate with spread-spectrum CDMA signals.

Cellular telephone from space is another promising possibility being actively pursued by at least a dozen different companies. The concepts they espouse range from the distribution of digital data through geosynchronous satellites to the use of large swarms of satellites swinging through space in low-altitude earth-skimming orbits. The many and varied characteristics of these various concepts will be described in detail in chapters to follow. But first, we will discuss conventional communication satellites and the spaceborne mobile communication systems now providing service to ships, long-haul trucking firms, and commercial transport planes.

## References

1. Neil J. Boucher, *Cellular Radio Handbook: A Reference for Cellular System Operation,* rev. ed., Quantum Publishing Co., Mendocino, Calif., 1992.
2. Ibid.
3. Ibid.
4. Andrew Kupfer, "Look Ma, No Wires," *Fortune,* December 13, 1993, pp. 147–154.
5. Ibid.

# 2

# Emerging Trends in Communication Satellites

*The Americans may have need of the telephone, but
we do not. We have plenty of messenger boys.*

SIR WILLIAM PREECE
*Chief of the British Postal System, 1876*

In the early days of the American space program, America's scientists and technicians launched artificial satellites that were almost embarrassingly small. Vanguard I—which Nikita Khruschev condescendingly nicknamed "the beeping grapefruit"—weighed only 3.25 lb.

America's inexperienced rocketeers blasted the Vanguard into orbit in 1958, when dollar bills were far more valuable than they are today. But, even if we measure Vanguard's cost in the largely uninflated greenbacks of the late 1950s, building and launching that tiny little satellite cost about $350,000/lb, or about 20 times the per-pound rate charged for today's far more capable family of orbiting satellites.

In the intervening years, of course, our biggest satellites have grown to gargantuan proportions. Some of them are bulkier than the long-haul trailer trucks whizzing along the New Jersey turnpike. A few military and civilian models tip the scales at 30,000 lb, or even more.

## Building a Thriving Business Along the Space Frontier

During the 35 years during which artificial satellites have been growing at such a rapid rate, space exploration has quietly become big business. The budget of the National Aeronautics and Space Administration (NASA) is today about $15 billion per year with a nearly equal amount being devoted to the military exploitation of space. Revenues from private, profit-making spaceborne ventures add another $20 billion or so per year to that total.

## The Three Beneficial Properties of Outer Space

The practical, economic benefits of the space frontier arise from three special environmental properties that come automatically when we journey into space: zero gravity, hard vacuum, wide-angle view. With modern technology, these three environmental properties are being exploited to the enormous benefit of the people living here on earth.

Hurtling along in the zero gravity of outer space, a space shuttle astronaut is able to gulp floating shrimp from midair like a giant silver fish. This is possible because astronaut, shuttle, and shrimp are all being equally influenced by the same gravitational force. Consequently, all three of them experience weightlessness "free-fall" as they travel around the earth. The tiny g-forces felt by the astronauts are created by drag with the few whisper-thin molecules of air and the almost imperceptible radiation pressure caused by the sun shining on the shuttle. These tiny drag forces—which amount to only about one-millionth of a g—gently slow the shuttle orbiter, but they do not affect astronaut or shrimp inside.*

In the next few years the weightlessness or zero gravity—so enjoyed by today's orbiting astronauts—will provide a number of practical benefits to the people living on earth. Improved pharmaceuticals, new optical lenses, and better computer chips will all likely result from the weightlessness of space. So will new methods of entertainment, new metal alloys, and novel construction techniques.

The vacuum of space creates problems for aerospace designers, but it can also be used in economically important ways. For several decades, technologically oriented companies like General Electric and Radio Corporation of America (RCA) have profited handsomely from vacuum technology. Every incandescent light bulb houses a low-level vacuum; so does the picture tube in every color television set. Hard vacuums are also necessary in refrigeration equipment, in the freeze-drying of foods, and in the welding of certain dissimilar metals such as aluminum and molybdenum.

The virtual absence of air in space has allowed aerospace engineers to develop some remarkably inexpensive inflatable structures. In the late 1950s giant Echo balloons were inflated a few hundred miles above the earth. They reflected radio messages from one ground transmitter to another. In one early experiment, a crude television image was bounced from coast to coast. It originated in Boston and it read "M-I-T." Inflation of those enormous silver bubbles was possible only because of the natural vacuum of outer space. On the ground, working against atmospheric pressure, 800,000 lb of gases were required to inflate an Echo balloon. But in the vacuum of space, only about 30 lb of gases were needed.

The wide-angle view from space is also providing practical benefits to you and your fellow earthlings, too. Fly on a commercial jetliner at a cruising altitude of 37,000 ft and your horizon will be about 200 nautical mi away. At higher altitudes, much more of the earth comes into view. At 100 nautical mi the astro-

---

*One g is defined as the acceleration due to gravity we experience on the surface of the earth.

nauts can see about 7 percent of the world at any given instant. At 19,300 nautical mi, one favorite vantage point for communication satellites, electromagnetic transmission links gain instantaneous access to 42 percent of the globe.

Communication satellites are today's most profitable application of the wide-angle view from space. Annual revenues currently exceed $16 billion, with compound growth rates of 20 to 30 percent. Twenty-four countries, including Indonesia and Brazil, own and operate their own communication satellites, and, at last count, 170 nations had ground antennas capable of receiving messages from space. While you are reading this paragraph, more than 100,000 telephone calls are passing through communication satellites. So are several dozen color television shows. Religious sermons, video conferences, computer data, and stock-market quotes all depend on modern satellite transmissions.

When the first Sputnik swept across the sky above the North American continent, one irate congressman argued that we should use our rockets to blast it to smithereens. He reasoned that it was illegally intruding into America's air space. Today we are lucky we did not choose to extend our air-space rights upward into space because the industrialization of the space frontier has become a multibillion dollar enterprise largely dominated by the United States.

## The Growing Importance of Spaceborne Telecommunication Services

In the decades ahead, millions of white-collar professionals and production line workers alike will be trained to do their jobs more efficiently by enormous numbers of words and pictures flowing through outer space, and direct-broadcast television shows by the dozens will soon be arriving at your doorstep from orbiting satellites. So will first-run movies and the latest video games.

Already today, whole newspapers are streaming through orbiting satellites so they will arrive at the same time all over the country. *The Wall Street Journal* and *USA Today,* for instance, are printed simultaneously at numerous remote printing sites connected together by satellites.

Other large news organizations are seriously studying new ways to deliver their printed stories directly into the home with tailor-made editions targeted toward specific individuals or small, tightly focused groups. This is how journalist Jonathan Weber describes current production methods for his newspaper, *The Los Angeles Times:* "The $230 million Olympic Boulevard press building, which covers nearly 1 million square feet, turns 430 tons of newsprint and 700 gallons of ink into 600,000 newspapers every day—newspapers that are then delivered across the southland by hundreds of trucks."[1]

Using a combination of satellites, computers, laser printers, and the like, Weber and his colleagues are hoping to find a better way to get their stories into the hands of selected readers nationwide.

Portions of the "want ads" are already being relayed through outer space. Orbiting satellites are routinely pressed into service to sell used cars in a nationwide dealer hookup that easily pays for itself in reduced travel costs alone. Aucnet USA of La Jolla, California, allows dealers to attend auto auctions while sipping coffee in their own private offices. Photos and descriptive information concerning the vehicles ricochet through the Galaxy 7 satellite to dealerships scattered around the countryside. Last year nearly 135,000 cars and trucks found new owners among the 3000 dealers, who paid $500 each to gain access to the electronic matchmaking capabilities of the Galaxy 7.

Electronic teleconferencing via satellite has also become big business, so big the airlines and aircraft manufacturers have commissioned studies to find out how much electronic meetings will likely dent their business in future years. Some experts are even predicting that future commercial jetliners will be rigged with teleconferencing equipment so the busy business traveler can attend one meeting electronically while he is hurtling toward another.

### Revenue Projections for Century 21

According to a study conducted by the Panel on Telecommunications at Loyola College in Baltimore, Maryland, total revenues from nonmilitary spaceborne communication services reached a total of $16.4 billion in 1992. The study authors also projected $43.3 billion in annual revenues for a decade later in 2002.

Some specific segments of the spaceborne telecommunications market are slated to grow at an even faster pace (Fig. 2.1). In particular, mobile communication systems based in space are expected to expand by a factor of 12 during that same decade, and video broadcast satellite markets are slated for a 16-fold growth level improvement.

## Historical Perspectives

So far, two basic types of communication satellites have been boosted into space: *passive satellites* which reflect radio signals back down toward earth, and *active satellites* which retransmit the signals they receive.

### Passive Communication Satellites

In its day, the Echo balloon was the world's best-known passive communication satellite, in part because it could be seen with the naked eye from the ground. Echo I was a 100-ft sphere of Mylar coated on the outside with aluminum. Mylar is an amazingly strong space-age plastic used commercially to make expensive tape recording tape. Echo I was launched in 1960 into an 870-nautical mi orbit inclined 47.3 degrees with respect to the equator.

The original Echo balloon gradually lost its reflectivity while it was coasting through space. Once the inflating gases leaked out, it became prune-faced and reflected radio waves inefficiently. A later version, Echo II, was semirigidized, that is, it was designed to remain spherical in space even after it lost all of its gases. This was accomplished by coating it on the outside and the inside with layers of aluminum and overstressing the Mylar inside the sandwich during its inflation. Like prestressed concrete, it then maintained its shape against compressive loads. Echo II was 135 ft in diameter. The sketches in Fig. 2.2 show how it was inflated on the ground and in outer space.

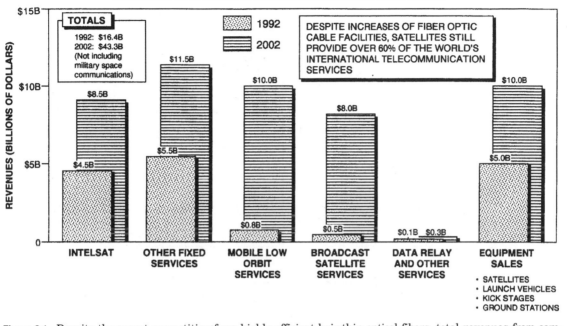

**Figure 2.1** Despite the recent competition from highly efficient hair-thin optical fibers, total revenues from communication satellites amounted to $16.4 billion in 1992. By the year 2002, industry experts are expecting this total to grow to $43.3 billion as larger numbers of advanced communication satellites are lofted into space. All market segments will grow expansively, but the two with the fastest growth rates will probably turn out to be mobile communication constellations in moderate-altitude orbits and video broadcast satellites at geosync. (*Source: Burton Edelson, Joseph Pelton, and Neil Helm, "Panel on Telecommunications," JTEC/WTEC, 1993.*)

A far more imaginative passive communication satellite, called Project West Ford, consisted of 1.2 billion dipole needles scattered along a circular orbit (Fig. 2.3). The dipoles were designed to reflect radio signals of a specific frequency like a thick swarm of resonant tuning forks. On hearing of the project, some of the world's leading astronomers protested. They were afraid that the ¾-in. needles floating in space would interfere with optical observations, pose hazards to future space travelers, and jam radio telescopes. But, nevertheless, the needles were blasted into space. Unfortunately, their mechanical dispenser apparently failed to work, and they were deployed in clumps rather than individually. That theory was never verified because no one was able to find the needles. A second launch in 1963 produced a properly deployed belt of needles which interfered with radio astronomy to some extent.

Some design engineers advocated making the needles out of beta tin, because, after a few hundred day-night cycles, beta tin disintegrates into a harmless gray powder. However, mission planners eventually decided to use copper wires instead, and let solar radiation pressure slowly shove them back into the earth's atmosphere. The thin belt of West Ford needles was used in dozens of communication experiments over a 3-year period before they began to come back.

## INFLATING THE ECHO BALLOON

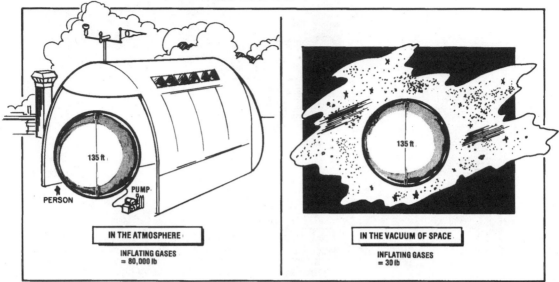

**Figure 2.2**  The vacuum of space helped simplify the design and inflation of the Echo balloon, which served as a passive communications satellite. In the early 1960s when test engineers inflated the Echo balloon in an old New Jersey dirigible hanger, they needed 800,000 lb of gases to overcome the crushing pressure of the atmosphere, but, when they inflated that same balloon in space, only 30 lb of inflating gases were required. The Echo balloon was produced in two different models. Echo I was 100 ft in diameter with a single coating of aluminum on its outer surface. Echo II was 135 ft in diameter with aluminum coatings on both its outer and inner surfaces.

Unfortunately, passive communication satellites were not a particularly efficient way to transfer large quantities of information from one ground station to another. Active satellites, crammed with electronic devices, it turned out, provided a far better approach.

### Active Communication Satellites

Active communication satellites are "repeaters"—they pick up faint radio signals from the ground, amplify them electronically, then retransmit them on a different frequency.

America's first active communication experiment was conducted in conjunction with Project SCORE on December 18, 1958. The satellite was launched into an elliptical orbit with a 97-nautical mi perigee and an 804-nautical mi apogee. SCORE's 8-watt transmitter relayed messages in real time, but the satellite was also rigged with store-and-forward messaging capabilities. As it coasted around its orbit, the SCORE satellite repeatedly rebroadcast a tape-recorded "Christmas greeting" message from President Eisenhower to the people in other countries of the world.

Four years later in 1962, Telstar, a talky electronic beach ball designed and built by AT&T, was lofted into space. Telstar skimmed over the earth in a low-altitude egg-shaped orbit that carried it from horizon to horizon in 60 minutes or less. Tracking antennas on the ground had to twist and turn in complicated patterns to remain properly aligned with Telstar, whose orbital motion was combined with the motion of the rotating earth to produce a wiggling trajectory across the sky.

Telstar was a lopsided sphere 32 in. in diameter weighing 172 lb. Its outer surface was covered with silicon solar cells to generate electric power and whiskerlike antennas to relay radio messages back down to the ground. Its maximum transmitter power was barely 2 watt, so large ground-based anten-

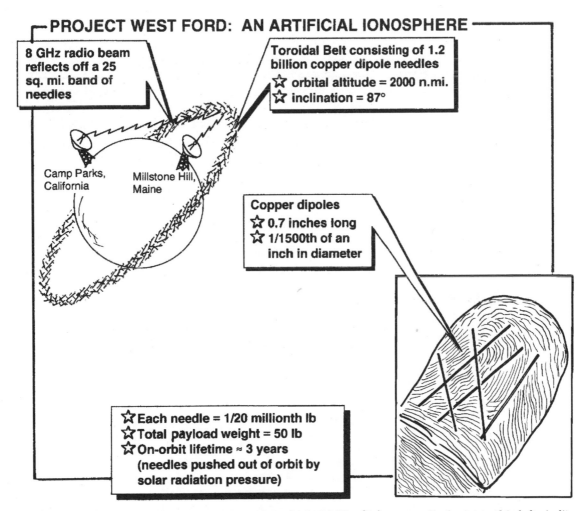

## PROJECT WEST FORD: AN ARTIFICIAL IONOSPHERE

**8 GHz radio beam reflects off a 25 sq. mi. band of needles**

**Toroidal Belt consisting of 1.2 billion copper dipole needles**
☆ orbital altitude = 2000 n.mi.
☆ inclination = 87°

Camp Parks, California

Millstone Hill, Maine

**Copper dipoles**
☆ 0.7 inches long
☆ 1/1500th of an inch in diameter

☆ Each needle = 1/20 millionth lb
☆ Total payload weight = 50 lb
☆ On-orbit lifetime ≈ 3 years
(needles pushed out of orbit by solar radiation pressure)

**Figure 2.3**  In 1963, a team of aerospace engineers launched 1.2 billion ¾-in. copper dipoles into a thin belt circling around the earth at an altitude of 2000 nautical mi. Once the dipole "needles" were properly dispersed, they served as an artificial ionosphere reflecting 8-GHz radio waves between pairs of ground-based transceivers at widely separated locations. The needles are gone now because solar radiation pressure caused them to reenter the earth's atmosphere.

nas were required to pick up its faint radio signals. The 85-ft antenna at Andover, Maine and a similar one at Pleumeur-Bodou in France were used repeatedly for message-relay purposes.

Telstar demonstrated that voice and video images could be relayed successfully through outer space, but its low power levels, nondirectional antennas, and irregular motion across the sky were not conducive to the transfer of wideband signals like the ones handled by today's far more capable communication satellites.

## Communication Satellites at Geosync

Most of today's communication satellites are boosted into "geosynchronous" orbits 19,300 nautical mi above the earth's equator. At that particular altitude, the angular speed of a satellite exactly matches the earth's rotational rate (one revolution per day). To earthlings standing on the spinning earth, such a satellite seems to hang ghostlike and motionless in the sky. A communication satellite at geosync is like a radio relay station mounted on an invisible tower, a tower so tall the satellite can "see" 42 percent of the earth—compared with 0.003 percent for a transmitter situated on top of the Empire State Building.

Except for two tight-fitting skullcaps near the north and south poles, three satellites spaced along the geosynchronous arc could theoretically serve the communication needs of the entire globe (Fig. 2.4). However, communication loads have become so heavy, that much larger numbers of satellites are actually required to handle global communications. Today about 300 geosynchronous satellites—most of them relaying commercial or military communications—form a thin metal necklace around the earth.

How does a geosynchronous satellite manage to hover in the sky? If you tie your shoestring to a stone and whirl it around in the air, it will follow a circular path similar to the orbit of a satellite. Like the stone, a satellite is held in orbit—against the pull of gravity—by its high rate of speed. A satellite just skimming over the atmosphere must be moving at a speed of about 17,000 mi/h to remain in orbit. It travels all the way around the world in about 90 minutes.

Gravity gets weaker at higher altitudes. So a satellite at that level can travel slower and still remain in orbit. Our moon orbits at an altitude of 208,000 nautical mi. Its speed is only about 2000 mi/h so it takes 28 days to complete one circuit around the earth.

A geosynchronous satellite orbits between these two extremes at a 19,300-nautical mi altitude. It travels at a bit less than 7000 mi/h and completes an orbit in 24 hours. Thus, if it is positioned above the equator, it will appear to hang motionless as viewed from the spinning earth.

Signals sent up from the ground to a geosynchronous satellite are usually clustered into batches containing several hundred two-way phone conversations to be picked up and rebroadcast by a single satellite transponder. A modern satellite is equipped with 24 active transponders, or even more, giving it a capacity of several thousand simultaneous conversations plus several channels of high-quality color television.

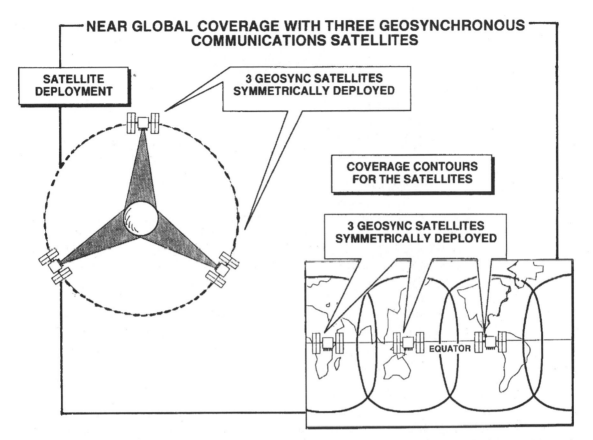

**Figure 2.4** Three properly spaced geosynchronous satellites launched into circular, equatorial orbits 19,300 nautical mi above the equator could theoretically provide continuous communication coverage for nearly all of the heavily populated portions of the earth. As the sketch at the bottom of this figure indicates, their coverage circles overlap one another to create a band stretching above and below the equator by ±50 degrees. Although three satellites would theoretically suffice, spacefaring nations have actually orbited 300 functioning geosynchronous satellites.

The radio waves used by communication satellites travel at the speed of light, but the geosynchronous altitude is so high it takes each message a quarter of a second to make the round trip up to the satellite and back down to the ground again. Most business callers eventually learn to ignore the delay, but, because of the way terrestrial telephone circuits are designed and clustered, each spoken word tends to produce an audible echo. Special echo suppressors can be used to damp out the echo, but sometimes faint traces still remain.

Communication specialists are constantly devising more economical ways to relay electromagnetic transmissions through outer space. Their contributions have included stronger structural materials, more efficient solar cells, and improved schemes for message switching. For many years they have become intrigued with *complexity inversion,* a special concept whereby designers purposely make their satellites large and complicated—and, consequently, much more expensive—so that they will be able to communicate with small, simple, widely proliferated, and inexpensive equipment on the ground.

This trend is clearly exemplified by two early American satellites. The first Telstar was a 172-lb bantam weight with tiny whip antennas capable of clear communications only with earth-based gargantuans such as the 85-ft dish at Goldstone, California. Less than a decade later, the ATS-6 weighing 3000 lb was launched into space. Its 30-ft dish-shaped antenna could carry on intelligible conversations with hundreds of 9-ft antennas scattered around the globe. Complexity inversion has reached its full potential now that the spacefaring nations of the world have begun to launch spaceborne systems capable of communicating with pocket telephones equipped with antennas only a few inches long powered by small disposable or rechargeable batteries.

Most early communication satellites were shaped like giant snare drums rolling through space. Their curved sides were populated with thousands of solar cells capable of picking up energy from the sun to power their radio transmitters. Like a toy gyroscope, such a satellite maintains stability because of its high rate of spin. Mechanical or electronic devices are used to despin the antennas so that they are always oriented toward earth.

More recently, nonrotating three-axis stabilized satellites have been introduced. Heavy flywheels inside a nonrotating satellite and tiny rockets mounted on the outside keep its main body and its communication antennas pointed toward specific spots on earth. Automatic mechanisms guided by infrared sensors continuously swivel its winglike solar arrays toward the sun. This approach (another clear-cut example of complexity inversion) is much more complicated and expensive than spin stabilization, but, especially for large satellites, it can provide more transmitter power per pound of spacecraft and more flexible communications.

## Early Bird and the Intelsat Series

Today's Intelsat satellites provide international voice, data, and television transmissions for nearly all of the nations on the surface of the globe. The earliest one, Early Bird, weighed only 85 lb. Early Bird, which was also called Intelsat 1, beamed a 6-watt signal down toward earth. Once it reached its destination in space, Early Bird provided 240 simultaneous two-way voice circuits, or, alternately, one low-quality television channel. Early Bird's C-band communication link used a 6-GHz uplink frequency and 4 GHz for the corresponding downlink. On April 6, 1965, it was blasted into its geosynchronous orbit. In the meantime, seven generations of Intelsat satellites have been built by Hughes and Ford Aerospace (now Loral), each far more capable than any of its predecessors. A typical satellite from the series, Intelsat 5, is sketched in Fig. 2.5, which also features a revealing drawing highlighting the population explosion along the geosynchronous arc.

The Communications Act of 1962 paved the way for the establishment of Intelsat, a well-organized international telecommunications consortium with headquarters in Washington, D.C. Comsat Corporation, which is partly owned

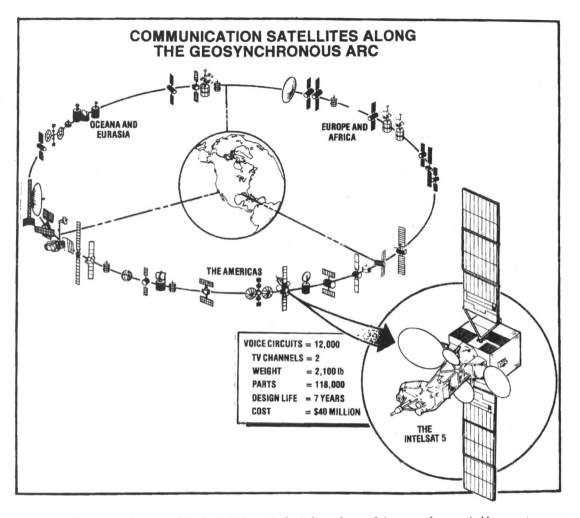

**Figure 2.5**  The geosynchronous altitude 19,300 nautical mi above the earth is presently occupied by a vast swarm of civilian and military satellites built and launched by many nations. These invaluable machines handle most of today's international telephone conversations and television transmissions. The Intelsat 5 is typical of this modern breed of communication satellites. It can handle 12,000 simultaneous telephone calls plus two channels of high-quality color television. Intelsat 6 and Intelsat 7 are even more capable communication satellites. Each of them can handle at least 40,000 simultaneous conversations.

by Wall Street stockholders, is the American entity that belongs to Intelsat. A sister consortium called Inmarsat uses separate satellites to provide similar communication services for large vessels at sea. Headquarter offices for Inmarsat are located in London, England.

By taking advantage of advancing technology, Intelsat's design engineers have managed to reduce the cost of spaceborne communication by several orders of magnitude. Hardware investment costs per satellite voice circuit over the past 30 years are depicted in Fig. 2.6. When Early Bird reached orbit in

## EMERGING COST TRENDS FOR COMMUNICATION SATELLITES

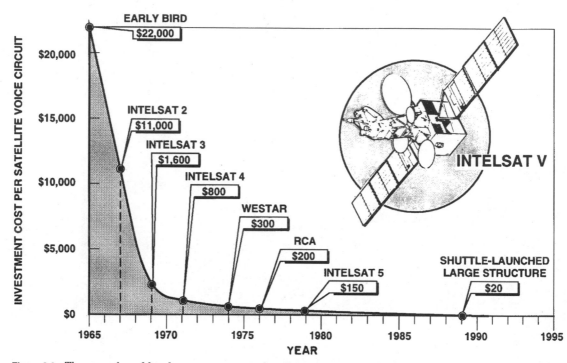

**Figure 2.6**   The space-based hardware necessary to handle an international telephone call costs only 1 percent of what it did 30 years ago. In 1966, the hardware for an Early Bird duplex voice circuit cost more than $20,000 per year. Today's satellites provide the same service with hardware that costs only about $150, and some experts foresee costs as low as $10 per year when large shuttle-launched antennas are eventually assembled in space. As costs drop, new communication services such as electronic teleconferencing and direct-broadcast television that would have been unthinkable 30 years ago are being marketed by private, profit-making companies, national postal systems, and international consortiums.

1965, each two-way voice circuit had an associated hardware cost of $22,000. By 1980, the corresponding cost had dropped to only about $150.

Moreover, in future years, hardware costs may drop into the $10 range for satellite communication links with the same basic capabilities. If and when this happens, overall hardware costs for the same service will have dropped by a factor of 2000 in half a lifetime.

## Geosynchronous Spacecraft Design

When aerospace engineers design a new satellite bound for outer space, they break it down into separate modules or subsystems, each devoted to a separate function. For a large, complicated communication satellite, such as the ones now hovering along the geosynchronous arc, an autonomous design team is usually assigned to work on each separate subsystem.

For modern communication satellites, the most important mission-unique subsystems are:

1. Orbit injection subsystem

2. Electrical power

3. Altitude and velocity control

4. Communication subsystem

These four key spacecraft subsystems are discussed one by one in the next four subsections.

## The Orbit Injection Subsystem

The booster rocket for the Syncom II communication satellite delivered its payload into a 100-nautical mi parking orbit with an orbital inclination of 28.5 degrees. The 28.5-degree inclination of the parking orbit resulted from the fact that its booster rocket was launched due east out of Cape Kennedy, which has a latitude of 28.5 degrees.

After it coasted southward to its equatorial crossing point, a perigee kick-motor was fired parallel to the surface of the earth. This increased the speed of Syncom II, thus driving it onto a transfer ellipse whose apogee (highest point) was equal to 19,300 nautical mi—the geosynchronous altitude. A second coast interval carried it up to apogee where a horizontal apogee kick-motor burn circularized the satellite's orbit at geosync.

This two-burn method of transfer from one circular orbit to another using horizontal rocket burns was developed by the German researcher, Hohmann, in 1925. Not surprisingly, it is called the Hohmann Transfer Maneuver.

Unfortunately, after circularization, Syncom II ended up in an *inclined* geosynchronous orbit with an inclination angle equal to 28.5 degrees. So, instead of remaining stationary over a specific point on the earth's equator, Syncom II traced out a "figure eight" ground trace as indicated by the small sketch on the left side of Fig. 2.7. The period of the figure eight ground trace was 24 hours, after which it traced out exactly the same figure eight pattern all over again.

The figure eight ground trace of an inclined geosynchronous satellite, such as Syncom II, results from a combination of its circular orbital motion and the relentless rotary motion of the spinning earth.

The sine-curve-shaped contour in Fig. 2.7 shows what the satellite's ground trace would look like on a Mercator projection if the earth did not rotate. It would, of course, trace out the same sine curve over and over again every 24 hours. The earth does, however, rotate on its axis. For every hour that elapses, it moves out from under the satellite's orbit 15 degrees in the eastward direction. The six horizontal stripes running across the figure show how the earth's rotation gradually moves the sine curve westward. The alternate gray and black stripes represent exactly 2 hours of travel for the earth (30 degrees). Notice how the two combined motions produce a figure eight ground trace.

## ROCKET PROPULSION FUNDAMENTALS

White-hot combustion by-products blasted rearward with blinding speed generate the rocket's propulsive force that hurls a rocket skyward. Pressure inside the rocket combustion chamber pushes in all directions to form balanced pairs of opposing forces which nullify one another, except where the hole for the exhaust nozzle is placed. Here the pressure escapes, causing an unbalanced force at the opposite side of the combustion chamber that pushes the rocket up toward its orbital destination.

Both rockets and jets are based on the same principle that causes a toy balloon, carelessly released, to swing in kamikaze spirals around the dining room. A jet sucks its oxygen from the surrounding air, but a rocket carries its own supply of oxidizer onboard. This oxidizer can be stored in a separate tank, mixed with the fuel, or chemically embedded in oxygen-rich compounds.

A *liquid* rocket usually has two separate tanks, one containing the fuel, the other containing the oxidizer. The two fluids are pumped or pushed under pressure into a small combustion chamber above the exhaust nozzle, where burning takes place to create thrust.

A *solid* rocket is like a slender tube filled with gunpowder; the fuel and oxidizer are mixed together in a rubbery cylindrical slug called the grain. Solid propellants are not pumped into a separate combustion chamber. Instead, burning takes place along the entire length of the cylinder. Consequently, the tank walls must be built strong enough to withstand the combustion pressure.

Rocket design decisions are dominated by the desire to produce the maximum possible velocity when the propellants are burned. A rocket's velocity can be increased in two principal ways: by using propellants with a high efficiency and by making the rocket casing and its engines as light as design constraints permit.

Unfortunately, efficient propellants tend to have some rather undesirable physical and chemical properties. Liquid oxygen is a good oxidizer, but it will freeze all lubricants and crack most seals. Hydrogen is a good fuel, but it can spark devastating explosions. Fluorine is even better, but it is so reactive it can actually cause metals to burn.

Miniaturized components, special fabrication techniques, and high-strength alloys can all be used to shave excess weight. But there are limits beyond which further weight reductions are impractical. The solution is to use "staging" techniques whereby a series of progressively smaller rockets are stacked one atop the other. Such a multistage rocket cuts down its own weight as it flies along by discarding empty tanks and heavy engines. However, orbiting even a small payload with a multistage rocket requires an enormous booster. The Saturn V moon rocket, for example, outweighed the Apollo capsule it carried into space by a factor of 60 to 1.

Syncom II's design engineers purposely lofted their satellite into an *inclined* geosynchronous orbit because their kick-stages did not have enough performance to make the necessary plane changes to achieve a 0 degree orbital inclination. But today's kick-stages are designed to provide ample plane-change maneuvers so the satellites they launch go into geostationary equatorial orbits.

The necessary plane-change maneuvers are combined with the perigee and apogee maneuvers so a (slightly more complicated) two-burn Hohmann Transfer Maneuver is still executed. During the perigee burn the spacecraft is swiveled to the side the proper number of degrees to combine the plane-change maneuver and the apogee-raising maneuver. The apogee kick-burn also involves a canted orientation, but it is cocked off at a much greater angle. For simplicity in some missions, the entire plane-change maneuver is made at the apogee of the transfer ellipse. However, this approach entails a performance penalty of about 80 ft/s compared with a maneuver sequence that uses an optimal plane-change split.

## COMPONENTS OF MOTION THAT MAKE UP THE FIGURE 8
## GROUND TRACE FOR AN INCLINED GEOSYNCHRONOUS ORBIT

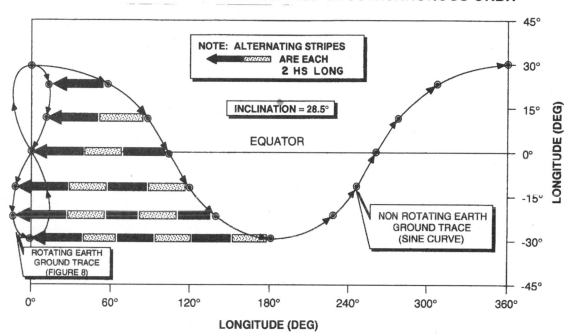

**Figure 2.7** A geosynchronous satellite launched into an inclined, circular orbit 19,300 nautical mi above the earth will trace out the same "figure eight" ground trace once each day. As this sketch indicates, the figure eight shape of the satellite's ground trace results from its movement around its circular orbit combined with the eastward rotation of the earth. With the earth's rotation frozen, an inclined geosynchronous satellite would trace out the sine-curve-shaped ground trace running across the map once each day. But, as the earth rotates, the points on the sine curve are moved to the left 15 degrees for every hour that elapses.

Some modern booster rockets can deliver their payloads directly onto a transfer ellipse bound for the geosynchronous altitude. In this case, the spacecraft needs only an apogee kick-motor, plus the usual trim-burn translational rockets to gently nudge it to its final destination along the geosynchronous arc.

Both solid and liquid rockets can be used to provide the perigee and apogee burns for geosynchronous satellites. Solid rockets are simple in design, safe, and easy to use, but the specific impulse (efficiency) of solid rockets is not as good as comparably designed liquid-fueled rockets.

A solid rocket is essentially a slender tube packed with gunpowder. A liquid rocket has been called a "plumber's nightmare."

Most liquid rockets have two separate tanks. One contains the fuel, the other contains the oxidizer. Upon ignition of the liquid rocket, the two fluids are pumped or forced under pressure into the combustion chamber where the fluid mixture is burned to produce thrust. The smaller trim burns are usually handled by hydrazine thrusters. Hydrazine is a liquid monopropellant contained in a single tank. Thrust is created by spraying the hydrazine over a catalyst which promotes a highly energetic chemical reaction. Hydrazine thrusters are also used for the periodic stationkeeping maneuvers necessary to keep the satellite within an acceptable distance of its assigned geosynchronous slot.

## Electrical Power Subsystem

Solar cells provide the electrical power used by geosynchronous communication satellites. They convert the sun's energy directly into electrical power. Solar cells are solid-state electronic devices first developed in 1954 for use in powering remote telephone systems.

Most solar cells consist of a thin layer of silicon properly doped with trace impurities to give its crystal lattice structure localized regions with excess electrons and holes—places where electrons ought to be, but are, in fact, absent.

Geosynchronous satellites occupy a fairly intense region of the upper Van Allen Radiation Belt where charged particles, primarily electrons, gradually damage their solar cells, thus reducing their ability to generate electrical power. Satellite designers oversize their solar arrays so they will still be providing enough electrical power at the end of their mission life.

Gallium-arsenide solar cells produce more electricity than silicon cells, and they better resist damage from the charged particles in the Van Allen Radiation Belts. But they are also, unfortunately, considerably more expensive. So silicon solar cells tend to be a more common choice.

A satellite at geosync spends about 1 percent of its on-orbit life in the earth's shadow. During those infrequent occulation intervals, the solar cells are unable to generate electrical power. To cover for the shadowing intervals and to provide for peak-load power demands, geosynchronous satellites are usually equipped with electrical storage batteries. Nickel-cadmium rechargeable batteries are the usual choice because they are fairly efficient, they can be hermetically sealed, and because no gas is involved during normal operation.

## Altitude and Velocity Control

Most modern geosynchronous communication satellites are shaped like giant cylindrical snare drums rolling through space. By spinning the main body of the spacecraft, designers can achieve gyroscopic stabilization, which helps maintain the proper attitude control for the satellite as it travels through space. For many satellites, the communication antennas must be despun—either mechanically or electronically—to maintain proper pointing toward earth.

The drum-shaped configuration yields a simple spacecraft design, but its generation of electrical power is not very efficient. Inefficiencies result from the fact that the solar cells, which are quite expensive, must be attached to the curving outer wall of the satellite's cylindrical body. As the body spins, different solar cells are illuminated by the sun's rays, but about half of the cells are, at all times, in the shadow cast by the satellite on itself.

An alternate design approach, called three-axis stabilization, puts the spinning mechanisms *inside* the spacecraft in the form of momentum wheels. Momentum wheels are heavy spinning flywheels usually mounted at angles with respect to one another in groups of three or four. By commanding electrical motors to modify the spin rates of the various flywheels, the spacecraft can continuously maintain the desired angular orientation due to the action-reaction principle enunciated by Isaac Newton.

Further fine-tuning for a spacecraft's attitude control systems is often accomplished by nutation dampers and sets of mutually perpendicular electromagnets. The magnets can be selectively activated as they cut across the earth's magnetic lines of flux, thus periodically slowing the rotation rates of the momentum wheels.

Attitude control thrusters, usually powered by hydrazine, also help with the attitude maneuvers for some orbiting satellites. The same supply of hydrazine may also power the translational thrusters that are used to maneuver the satellite to a new velocity and/or a new location in space.

Geosynchronous satellites, in particular, require frequent on-orbit maneuvers, because they must maintain such precise and accurate locations with respect to their neighbors along the geosynchronous arc. Perturbations due to the earth's nonspherical shape, the gravity of the moon and the sun, and solar radiation pressure have a big effect on a geosynchronous satellite's position in space. Depending on the mission, these various forces of perturbation may require frequent or infrequent compensating spacecraft maneuvers.

## Communication Subsystem

The primary purpose of a geosynchronous communication satellite is to relay modulated electromagnetic waves between widely separated ground-station antennas on earth. This is accomplished by picking up the modulated electromagnetic waves and immediately rebroadcasting them on a different frequency. Most modern geosynchronous satellites operate in three specific frequency bands set aside by regulatory bodies to handle spaceborne communications.

The most popular frequencies are in the C-band: 6 GHz up, 4 GHz down. More sophisticated users can, in addition, use the popular Ku-band frequencies: 13 GHz up, 11 GHz down. Gradually researchers are opening up the Ka-band in the vicinity of 30 GHz, although, in that electromagnetic regime, falling raindrops can cause fairly serious signal attenuation.

Telephone calls being relayed through a geosynchronous communication satellite are usually grouped together into batches ranging from 300 to 1000 calls. Each batch of calls is routed through an onboard receiver/transmitter called a *transponder*. The Intelsat 5 satellites are equipped with 24 transponders each. Intelsat 7 has 35 transponders. A transponder can also be used to handle a single channel of high-quality color television. For some satellites, specific transponders are dedicated to video; others make real-time tradeoffs between video and voice.

Most of the electronic processing and storage functions carried out onboard a satellite are handled by transistors and other types of solid-state semiconductors. But one vacuum tube sometimes still remains—the traveling-wave tube used to amplify the electromagnetic signals sent back toward earth.

As it makes its way through a traveling-wave tube, the electromagnetic signal follows a complicated spiraling trajectory. During that interval it is amplified by carefully controlled interactions with a beam of electrons swirling through the center of the spiral. A traveling-wave tube is efficient, compact, small, and light, and, even in heavy use, it will likely operate 5 years or more. However, simpler solid-state devices are gradually being adopted to handle the required amplifications.

## Overall Spacecraft Design

Aerospace designers must make numerous engineering tradeoffs as they attempt to find the best possible spacecraft design capable of handling a specific civil or military mission. Keeping the spacecraft weight down is nearly always a problem, and so is supplying ample quantities of onboard electrical power. Freezing the final design at just the right time can also turn out to be a tricky enterprise.

So far, America's aerospace engineers are preeminent in the design of geosynchronous communication satellites with a capture rate of about 75 percent of the world market. However, recent studies have shown that the rest of the world, especially the Europeans and the Japanese, are beginning to match our American expertise in some aerospace fields.

## Stationkeeping Techniques

To a first approximation, a man-made satellite launched into a circular orbit with a 0 degree orbital inclination at a 19,300-nautical mi altitude will remain stationary with respect to the spinning earth. In the real world, however, perturbations of various types disturb the satellite's orbit by pulling and pushing it around in space.

As Table 2.1 indicates, these forces of perturbation arise principally from three separate sources. Moreover, they cause three recognizable types of changes in the orbit of a geosynchronous satellite. Stationkeeping maneuvers are routinely used to mitigate the most worrisome effects of these perturbations. The sources and the major effects of the three principal types of perturbations acting on a geosynchronous orbit can be summarized as follows:

1. Earth's tesseral harmonic: apparent longitudinal (east-west) drift

2. Lunar and solar gravitational fields: orbital inclination perturbation

3. Solar radiation pressure: orbital eccentricity variations

In the next three subsections these three types of perturbations are discussed in detail together with discussions of some of their most important mission effects.

### East-West Longitudinal Drift

Cut a slice through the earth at the equator perpendicular to its north-south spin-axis, and the resulting cross section will turn out to be an ellipse, not a cir-

**TABLE 2.1  Systematic Perturbations Acting on the Orbit of a Geosynchronous Satellite**

| Source of the perturbation | Effect of the perturbation | Stationkeeping $\Delta V$ necessary to correct for the perturbation | Types of corrections usually made |
|---|---|---|---|
| Noncircular cross section of the earth's equator creates gravitational perturbations. | The satellite appears to drift eastward or westward along the equator. | The required velocity change varies with the satellite's assigned orbital position above the equator. The total $\Delta V$ is typically 7 ft/s per year or less. | East-west position corrections are usually made to avoid having the satellite intrude on the orbital slots of other, nearby satellites. |
| Gravitational perturbations from the sun and the moon distort the satellite's orbit. | The inclination of the satellite's orbit changes. This creates an increase or a decrease in the size of its figure eight ground trace. | Maintaining a 0 degree inclination requires a $\Delta V$ of 158 ft/s per year. | North-south corrections are usually made unless moderate inclination variations can be tolerated. The Comsat "nodding" maneuver can be used to minimize on-orbit propellant consumption. |
| Solar radiation pressure from the sun's rays systematically perturb the satellite's orbit. | The satellite's orbit experiences systematic changes in eccentricity. The eccentricity builds up for the first 6 months, then shrinks during the next 6 months. | The required velocity change for continuously maintaining a circular orbit varies with the satellite's ballistic parameter. Typically the total $\Delta V$ amounts to a few hundred ft/s per year. | Eccentricity corrections are usually unnecessary because small variations in the shape of the orbit can be tolerated for most missions. |

## THE INTELSAT 5 GEOSYNCHRONOUS COMMUNICATION SATELLITE

Imagine a mechanism 10 times as complicated as your family Chevrolet designed to operate flawlessly in a punishing environment for 7 full years—without servicing. Its name is Intelsat 5, fifth-generation descendant in a long line of distinguished machines. It is assembled from 118,000 separate parts, many of them flimsy and delicate, but redundant subsystems and clever design combine to give it a long lifetime of useful service.

The Intelsat 5 is a modern communication satellite. Built by Ford's Western Development Laboratories, it weighs 2100 lb and measures 51.5 ft wingtip to wingtip (the height of a five-story building). It retails for more than $40 million or about $20,000/lb—triple the current price of 24-karat gold. The Intelsat 5 turned out to be more capable and cost-effective than any of its forerunners. It was rigged to handle 12,000 telephone conversations plus two channels of high-quality color television.

Most communication satellites are shaped like huge snare drums rolling through space. Like a toy gyroscope, such a satellite remains stable because of its high rate of spin. Mechanical or electronic devices are used to "despin" its antennas so they are always oriented toward the earth. The Intelsat 5, however, does not rotate as it moves around its orbit. Its orientation is maintained by spinning flywheels inside and tiny rockets on the outside that keep its communication antennas in an earth-seeking orientation. Automatic mechanisms guided by infrared sensors continuously swivel its winglike solar arrays toward the sun. This approach is trickier and more expensive than spin stabilization, but for large satellites it can provide more transmitter power per pound of spacecraft and more flexible communications. However, despite the apparent advantage of the Intelsat 5's design and construction, its successor, the Intelsat 6, with three times more capacity, reverted back to the simpler snare drum design.

cle as you might expect. The major component of the earth's gravitational field caused by this elliptical cross section is called the "tesseral harmonic." The tesseral harmonic causes a satellite in a geosynchronous orbit to drift parallel to the equator in the eastward or the westward direction. The direction in which it moves depends on its initial longitude above the earth's equator. It will remain stationary if it happens to start out at one of four specific longitude locations.

Figure 2.8 pinpoints these stable longitudes. It also shows how much total velocity must be added each year to maintain the position of geosynchronous satellites assigned to various specific longitudes. A satellite positioned at the 225 degree east longitude, for instance, directly above the mouth of the Amazon River, would require stationkeeping maneuvers totaling 4.2 ft/s to stay at that position.

East-west stationkeeping is required for most geosynchronous satellites because they must not come too close to their neighbors in space. The primary difficulty with close encounters is not physical collisions, although such collisions are always a possibility. The primary difficulty is that, if two geosynchronous satellites come too close together, messages from the ground directed toward one of them may spill over onto its nearest neighbor in space.

As Fig. 2.8 indicates, the maximum change in velocity ($\Delta V$) required to make corrections for east-west longitudinal drift amounts to only about 7 ft/s per year even under worst-case conditions. A satellite that generates a velocity increment of 7 ft/s per year over a 7-year mission life will consume only about 0.75 percent of its weight in hydrazine propellants over that entire interval.

# ANNUAL EAST-WEST LONGITUDINAL STATIONKEEPING VELOCITY FOR GEOSYNCHRONOUS EQUATORIAL ORBITS

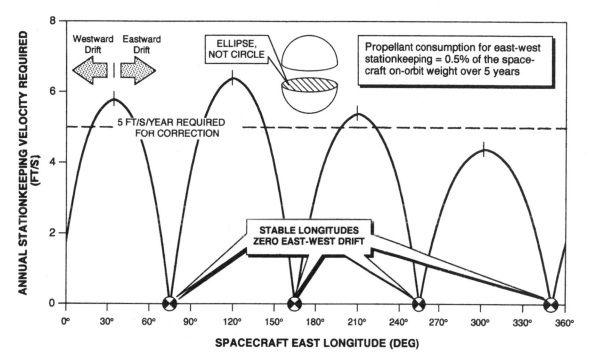

**Figure 2.8**  The cross section of the earth at the equator is an ellipse, not a circle. The corresponding distortion in the earth's gravitational field causes a geosynchronous satellite to begin drifting in the eastward or the westward direction, unless it happens to be located at one of the four stable points where these curves touch the horizontal axis. If the satellite is assigned an orbital slot at any other longitude, the stationkeeping velocity it must generate to stay there turns out to be, at most, 7 ft/s per year.

## North-South Drift of the Satellite

A satellite hovering over the equator at the geosynchronous altitude will experience systematic perturbations due to the relentless gravitational fields of sun and moon. If its orbit starts out with a 0 degree inclination, those gravitational perturbations will cause a gradual increase in its orbital inclination. If stationkeeping corrections are not made, the satellite will trace out an increasingly larger figure eight ground trace once per day over the life of its mission.

As the graph on the left-hand side of Fig. 2.9 indicates, a satellite that starts with a 0 degree orbital inclination will experience an inclination growth of about 0.8 degree per year. The small sketches in the figure show how the satellite's figure eight ground trace gets bigger and bigger with the passage of time.

North-south stationkeeping corrections are made for many of today's geosynchronous communication satellites so the parabolic dish antennas on the ground will not have to engage in satellite tracking throughout the day. North-south stationkeeping corrections of this type require velocity increments totaling 158 ft/s per year.

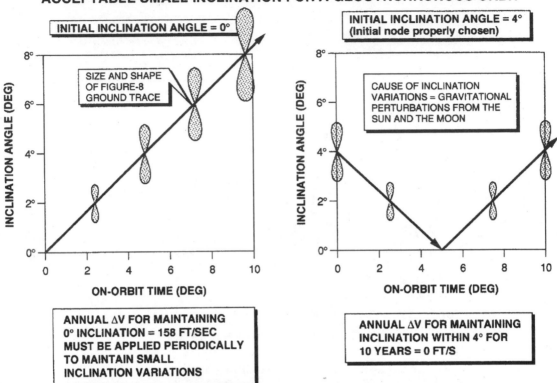

## NORTH-SOUTH STATIONKEEPING ΔV NEEDED TO MAINTAIN AN ACCEPTABLE SMALL INCLINATION FOR A GEOSYNCHRONOUS ORBIT

**Figure 2.9**  Gravitational attractions from the sun and moon cause gradual perturbations in the orbital inclination of a geosynchronous satellite. If the satellite is launched into a geosynchronous orbit with a 0 degree initial inclination, its inclination angle will build up at a rate of about 0.8 degree per year unless stationkeeping ΔV's totaling 158 ft/s per year are periodically applied. However, if an inclination of as much as 4 degrees can be tolerated over a 10-year mission, the inertial nodal crossing point of the satellite can be carefully chosen so its inclination starts at 4 degrees, declines to 0 degree after 5 years, then builds back up to 4 degrees again over the next 5 years.

Generating annual velocity increments that large puts a severe burden on a geosynchronous satellite in terms of propellant consumption. This limits its on-orbit life. Over a 7-year mission, for instance, a typical hydrazine-fueled satellite would consume 16 percent of its weight performing the required north-south stationkeeping maneuvers. When the satellite runs out of hydrazine, it can no longer maintain its orbital position.

In order to extend the lifetime of certain geosynchronous satellites, researchers at Comsat have perfected a so-called "nodding" maneuver that allows the satellite to perform communication functions while it is experiencing fairly large north-south orbital excursions. The spacecraft gradually swivels and tilts to compensate for its north-south motion. This so-called "Comsat maneuver" can double the useful mission life of some satellites, thus reducing the cost of communications services they provide. Comsat engineers have patented the nodding maneuver, and they willingly license the technology to other companies.

The Comsat maneuver works as advertised, but unfortunately, when it is being used, communication specialists on the ground must gradually swivel their antennas throughout the day to keep the satellite at the center of the ground-based antenna's field of view.

The graph on the right-hand side of Fig. 2.9 highlights another method for minimizing propellant consumption for missions in which moderately large north-south excursion can be tolerated. In this particular case, a 10-year mission is planned during which 4 degree north-south inclination excursions can be tolerated. To meet this goal without using any north-south stationkeeping propellants, the satellite is purposely launched into a geosynchronous orbit with a 4 degree inclination with its equatorial crossing points properly positioned with respect to the location of the sun and moon. This is achieved by scheduling liftoff for a particular time of day. Notice that the initial 4 degree inclination of the satellite's orbit begins to shrink instead of grow. During the first 5 years of its on-orbit lifetime, the satellite's inclination declines from 4 to 0 degrees. Then, during the last 5 years, its inclination builds back up to 4 degrees again.

## Eccentricity Variations

When the sun's rays illuminate any object, a small solar radiation pressure pushes on its surface. The resulting force amounts to only about 5 $lb/mi^2$, at most, but nevertheless the small force pushes so relentlessly over weeks and months, the resulting orbital perturbation can be quite noticeable.

The solar radiation pressure pushing on a geosynchronous satellite causes systematic variations in the eccentricity (oblateness) of its orbit. An orbit with an eccentricity of 0 is circular; one with an eccentricity close to 1 is greatly elongated. The apogee of a geosynchronous satellite whose orbit has an eccentricity of 0.01 will be 450 nautical mi higher than its perigee.

Throughout the year, the geometrical relationship between the sun and the orbital plane of a geosynchronous satellite systematically changes. The net result is that its orbital eccentricity increases during the first 6 months, then over the next 6 months shrinks back down to zero again.

Typical orbital eccentricity variations due to solar radiation pressure are depicted by the family of curves in Fig. 2.10. Each curve is labeled with a specific ballistic parameter. The ballistic parameter for a satellite equals its weight-to-area ratio divided by its drag coefficient. The drag coefficient is a dimensionless quantity. Usually for a high-speed vehicle moving through the virtual vacuum of space it is assumed to equal 2.0.

A 500-lb spacecraft with a cross-sectional area of 1000 $ft^2$ would have a ballistic parameter of 0.25. This corresponds to the value labeling the top curve of Fig. 2.10, which ends up with a peak eccentricity value of 1.8 percent (0.018). With a 1.8 percent eccentricity, a geosynchronous satellite's apogee altitude turns out to be 820 nautical mi higher than its perigee altitude. The ballistic parameter values for most of today's geosynchronous satellites are fairly large, so solar radiation pressure does not distort their eccentricities by a very large

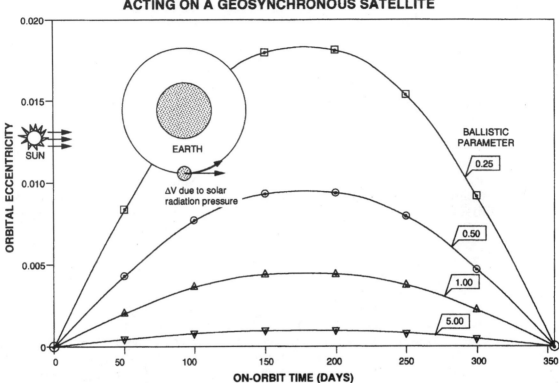

**Figure 2.10**  Solar radiation pressure exerted by the sun causes the eccentricity of a geosynchronous satellite to build up for 6 months, then decline to its original 0 degree value over the next 6 months before the process starts all over again. As these curves indicate, the total eccentricity variation depends on the satellite's ballistic parameter, which equals its weight-to-area ratio divided by the drag coefficient (usually 2.0). For a satellite with a ballistic parameter of 0.25, the maximum orbital eccentricity turns out to be 0.018.

amount. For most missions, aerospace engineers simply allow the orbital eccentricity to alternately build up and decline over the lifetime of the satellite.

Allowing the eccentricity to vary saves hydrazine propellants, thus lengthening the satellite's on-orbit life. Depending on the ballistic parameter of a particular satellite, the required $\Delta V$ to maintain its orbital eccentricity at a 0 value can amount to several hundred feet per second per annum.

## Overall Stationkeeping Strategies for Geosynchronous Satellites

The engineers and technicians who design and operate geosynchronous communication satellites are always on the lookout for ways to minimize any velocity increments the satellite will have to generate over its mission life.

Usually they cannot avoid east-west stationkeeping maneuvers because they must keep the satellite within a specific orbital slot and they must honor international agreements. Often they must perform north-south stationkeeping maneuvers to minimize tracking burdens on the ground antennas. Most of the time they can avoid stationkeeping maneuvers designed to eliminate eccentricity variations unless their mission is unusually demanding.

Fortunately, the on-orbit maneuvers required to eliminate the detrimental effects of tesseral harmonics, lunar and solar gravity, and solar radiation pressure are simple and easy to perform. Moreover, the required maneuvers can usually be delayed for many days without particularly worrisome consequences. For some missions certain other on-orbit perturbations such as materials' outgassing, higher-order gravitational harmonics, and nonsymmetrical thrusting and attitude adjustment maneuvers may also play a smaller role.

## Ground Station Design

Huge numbers of ground antennas are pointed up from the ground toward today's family of geosynchronous satellites. In the United States alone, for instance, nearly 3 million privately owned dish-shaped antennas are currently picking up cable television and other video signals from space. Many of them are owned and operated by hotels, motels, restaurants, bars, and private clubs. But others are owned by private citizens, often in areas where local television broadcasts and cable television services are readily available.

The biggest, most profitable, customer for today's domestic communication satellites is cable television. But television images of other types are also beginning to make substantial contributions in the overall marketplace. Sheraton Hotels, for instance, and several other major hotel chains receive their in-room movies from communication satellites. Airwave ministers, including Oral Roberts and Rex Humbard, are also heavy users of satellite transmissions.

Other big organizations are also beginning to make heavy use of orbiting satellites. The American Bar Association and the American Law Institute, for example, have signed an agreement with Comsat for a continuing legal education network. In a similar cooperative venture, J.C. Penney's department stores and the Private Satellite Network are installing 6-ft earth stations in hundreds of shopping malls nationwide. Among other things, the new network will distribute management training courses developed by the Penney chain. In addition, Penney's executives are marketing "video-conferencing services using shopping-mall motion picture theaters... gourmet cooking classes, and family financial planning instruction." Upscale ideas like this are helping to enhance the Penney image.[2]

Ground-based antennas are usually shaped like parabolas with the parabolic surface focusing the beam coming down from the satellite to direct its energy toward the feed mechanism located at the focus of the parabola. The feed mechanism routes the electromagnetic energy into the receiver, a solid-state device that processes the modulated waves into a usable format—voice, data, video, fax.

The received power in a downlink transmission is, to a first approximation, proportional to the *product* of the cross-sectional areas of the transmitting antenna on the satellite and the receiving antenna on the ground. If the two antennas were interchanged, the received power would be essentially the same. Of course, the received power is also directly proportional to the output power of the satellite, so enlarging the satellite's antenna or cranking up its output power can provide more power to the ground antenna. When more power comes down the ground antenna, it can be built smaller. Moving the satellite to a lower altitude, using spot-beam antennas with improved directionality, or using higher transmission frequencies can also help make ground antennas smaller—which is in keeping with the concept of *complexity inversion.*

Higher-frequency transmissions are gradually being opened up to sophisticated consumers around the world, to permit the use of smaller earth-station antennas. Unfortunately, with higher frequencies, pointing accuracy requirements also become more difficult to achieve, and the shape and surface figure of the antenna must be more precisely controlled, so it can properly reflect the higher-frequency electromagnetic waves.

Some antennas do not have parabolic shapes. Specifically, phased-array antennas, such as those being produced to pick up video signals from Europe's direct-broadcast satellites, are manufactured in the form of flat squares. Many of them measure less than 18 in. on a side.

## Orbital Overcrowding and Its Proposed Solutions

Approximately 300 communication satellites are presently positioned at the geosynchronous altitude, a circular arc 143,000 nautical mi long. On the average, these satellites are about 470 nautical mi apart, so it may seem ludicrous that mission planners would be worrying about the population explosion at geosync. But, in fact, if we don't soon implement preventive measures, tomorrow's communication experts may be faced with serious satellite overpopulation. Already, transmission frequencies are in short supply, and messages directed toward one satellite sometimes spill over onto its neighbors in space.

In the earliest days of our spacefaring era, "squatters rights" usually prevailed. When a constellation of satellites was lofted into space, it preempted specific locations and specific transmission frequencies. Naturally, shrill cries of protest came from Third World countries incapable of launching satellites of their own. Their objections reached a fever pitch if the satellite was hovering in the vicinity of their territory and, most especially, if their territory was on or near the equator.

Every few years these arguments are voiced anew at major frequency allocation conferences held at such exotic locations such as Geneva, Switzerland and Madrid, Spain. Third World representatives argue that the available orbital slots and transmission frequencies should be apportioned administratively once and for all. Delegates from the United States and western Europe maintain that premature allocations would waste precious resources and erect institutional barriers to innovative research. Some compromises have been achieved, but, by and large, Third World attitudes toward administrative apportionment have tended to prevail.

Fortunately, technical methods are also available for alleviating orbital overcrowding, some of which are listed in Fig. 2.11. Higher-frequency transmissions are helpful to some extent; a ground antenna of a given size sends out a

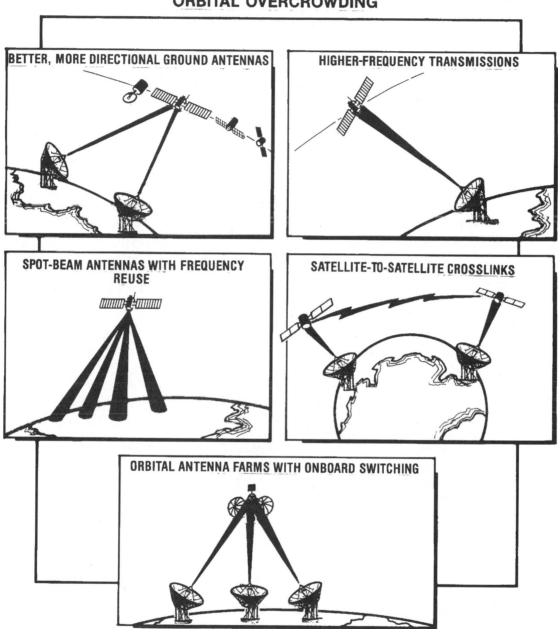

**Figure 2.11**  Scientific projections indicate that, if sufficient care is not taken, our growing population of geosynchronous communication satellites will eventually saturate the airwaves. Fortunately, various technical solutions—improved ground antennas and higher-frequency transmissions—will allow us to place increasingly larger numbers of satellites along the geosynchronous arc. In addition, multibeam antennas, orbital antenna farms, and the careful use of satellite-to-satellite crosslinks will help increase the capacity of each satellite so fewer of them will be required to handle our increasing traffic loads.

smaller beam if it is transmitting at a higher frequency. Early communication satellites could be positioned along the geosynchronous arc no closer than 3 to 4 degrees because they operated at relatively low frequencies in the 4-to 6-GHz range. More recently, those heavily used frequencies were doubled, to about 12 GHz, and some sophisticated users are doubling them once again to move into the 30-GHz regime.

Each jump in frequency requires expensive new equipment—in space and on the ground—but many more geosynchronous satellites can be positioned in space, in some cases, as close as 1 degree apart. Unfortunately, these higher-frequency transmissions send out shorter wavelengths that are absorbed by falling raindrops, a problem that sometimes dictates multiple receiving sites to achieve spatial diversity. Studies have shown that pairs of antennas on the ground can be spaced as close as 10 or 15 mi apart with a good chance of finding an open pathway to one ground station when heavy rainfall blocks the other.

Multibeam antennas carried aboard the satellites can be another fruitful way to conserve available frequencies. A multibeam antenna transmits a family of pencil-thin beams, often so small that, by the time they reach the ground, their footprint covers an oval only a few dozen miles wide. As a result, the same frequencies can be reused in several different parts of the world.

Polarized signals have also allowed frequency reuse on a grand scale. Like polarized sunglasses, the polarizing mechanisms on a communication satellite allow only vertically or horizontally polarized signals to get through. The resulting polarized beams can overlap on the ground in the same geographical region at the same frequency because, to a horizontally polarized antenna, a vertically polarized signal is essentially invisible.

Another conservation method, called "bandwidth compression," is based on the fact that, in most instances, only minor frame-to-frame changes occur in the successive pictures on a television screen. Hence it turns out to be more efficient to use digital processing techniques and transmit only those small portions of the picture that have changed since the last frame rather than transmitting a complete new picture. This technique produces acceptable-quality pictures with the equivalent of 60 voice circuits instead of the 600 normally required. Similar compression techniques allow more human voices to occupy the same portion of the electromagnetic frequency spectrum.

Satellite-to-satellite crosslinks, now passing into widespread use, conserve precious bandwidth because multiple hops to widely dispersed ground stations are no longer required. The Iridium mobile communication system, now being engineered, is slated to make extensive use of crosslinking techniques.

Multiple space shuttle launches with on-orbit satellite assembly could give rise to the next giant step in complexity inversion: orbital antenna farms. An orbital antenna farm is a large platform in space with a number of antennas each operating on a different frequency. Usually each separate set of antennas is devoted to a different service. Such devices could help reverse the trend toward the installation of increasing numbers of satellites at geosync. Some experts are convinced that users will lease portions of an orbital antenna farm while sharing electrical power, thermal control, and other utility services. This is similar to the way the shops in large shopping malls are now leased.

Between 1974 and 1977, as a key part of Rockwell International's space industrialization studies, my colleagues and I put together the plans for one of the world's first orbital antenna farms. It was a highly ambitious project involving a 65,000-lb geosynchronous platform almost twice as long as a football field. The platform was to be rigged with 30 antennas capable of performing five economically useful services for thousands of ordinary private citizens living on planet earth.

Engineering details of its construction and operation are presented in the next major section, which also highlights its five different communication services including pocket telephones to be supplied to hundreds of thousands of individual users at economically affordable rates.

## Early Design for an Orbital Antenna Farm

The design characteristics of Rockwell International's 500-kW orbital antenna farm are summarized in Fig. 2.12. Basically, it consists of a geosynchronous antenna platform, a power distribution unit, a 2-degree-of-freedom slip-ring structure, a gallium-aluminum-arsenide solar array, and a docking ring to allow shuttle on-orbit assembly in a low-altitude 28.5 degree parking orbit.

Its 500-kW solar arrays are attached to nine beams each of which is 584 ft long. The beams are manufactured using special on-orbit beam-making machines which also help attach the coil-wound solar-cell blankets. The gallium-aluminum-arsenide solar cells, which have a concentration ratio of 2:1, convert the sun's energy into electricity with a conversion efficiency of 18 percent. Total operational weight of the system including on-board consumables is approximately 65,000 lb.

Once the orbital antenna farm has been constructed in its low-altitude parking orbit, it is self-propelled to its final geosynchronous destination-orbit using its own solar electric power. The two solar arrays gather electricity to power the thrusters. When it is in position along the geosynchronous arc, it is maintained, repaired, and serviced by a dedicated, remotely controlled teleoperator robot, which remains permanently in orbit with the facility. Low-intensity laser beams relay digital pulse-trains to and from duplicate antenna farms performing similar services for Europe, Asia, and other parts of tomorrow's world.

The orbital platform is equipped with 30 different antennas capable of performing all five of the required communication services. The biggest antennas, which are 60 ft in diameter, are devoted to the pocket telephone communication links. The smallest antennas are devoted to electronic teleconferencing. Their diameter is just over 3 ft.

Most of the power produced by the unit, over 90 percent of it, is devoted to pocket telephones and direct-broadcast television. As Fig. 2.12 indicates, the platform is rigged with several different multibeam antennas that transmit families of pencil-thin beams toward users on the ground. The spot beams from

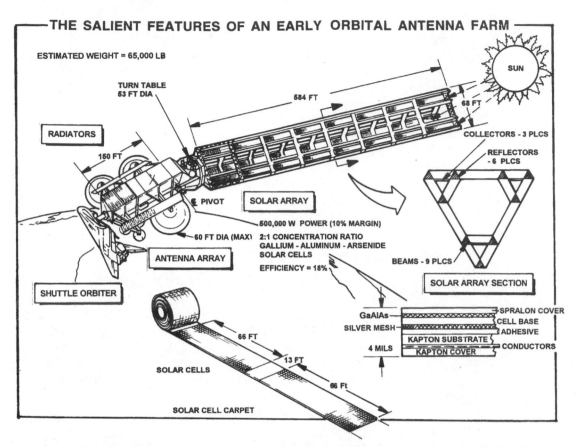

**Figure 2.12**  This orbital antenna farm, which was designed by the author and his colleagues at Rockwell International in the 1970s, was to be assembled in a low-altitude orbit, then flown under its own power to the geosynchronous altitude. Total on-orbit weight was 65,000 lb with enough gallium-aluminum-arsenide solar cells to generate 500,000 watts of electrical power. The antenna farm, which was equipped with 30 different multibeam antennas, was slated to perform five communication services: direct-broadcast television, mobile pocket telephones, national information services, electronic teleconferencing, and electronic mail.

the multibeam antennas provide for extensive frequency reuse in which the same frequencies are reused several times in widely separated geographical regions. This greatly reduces overall bandwidth requirements. The multibeam antenna approach also promotes a simpler and lighter design.

The 500-kW orbital antenna farm is designed to perform five different nationwide communication services:

1. Direct-broadcast television with five simultaneous color television channels operating 16 hours per day.

2. Pocket telephones with 45,000 duplex voice channels connecting shirt-pocket telephones to one another and to the ordinary phones in the conventional telephone networks. Assuming a 5 percent peak usage rate, 900,000 pocket telephones could be sold before expansion of the space links would be required.

3. National information services with direct access through the orbital antenna farm to libraries, computer data banks, human experts, and archival data stored on audio and video disks.

4. Electronic teleconferencing with the capability for two-way video links set up between as many as 300 widely separated ground sites. Each site is equipped with studio facilities, video cameras, computers, wide-screen monitors, switching gear, and separate audio-communication links.

5. Electronic mail with facsimile transmission of personal and business correspondence to 800 regional centers, interconnected through the orbital antenna farm. The capability goal calls for the delivery of 40 million pages of mail with overnight hard-copy delivery to the intended recipients.

The sketches in Fig. 2.13 show how the pocket telephones are connected with the existing telephone networks. The pocket telephones are rigged with 3-in.

## ARCHITECTURE FOR THE POCKET TELEPHONE SYSTEM TO BE INSTALLED ON BOARD THE ORBITAL ANTENNA FARM

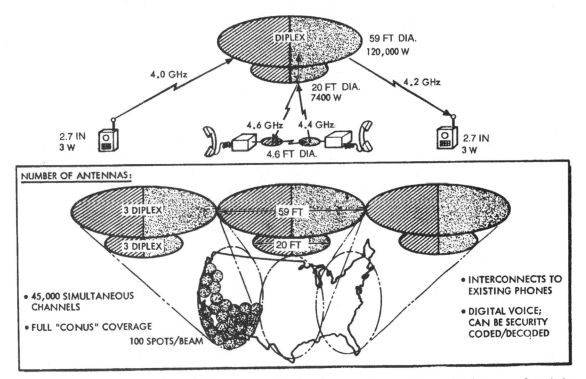

**Figure 2.13** These sketches show how the pocket telephone system carried aboard the orbital antenna farm is interlinked with the existing telephone networks. Three-watt hand-held transmitters equipped with 3-in. whip antennas transmit conversations from the ground directly to a 60-ft dish antenna on the satellite. This antenna, in turn, feeds the transmissions into the satellite's 20-ft parabolic dish for direct relay to the switching network located on the ground. One advantage of this approach is that it links the shirt-pocket telephones to all the ordinary telephones in the existing system. Another is that it allows the complicated switching devices to be located on the ground where maintenance and modifications can be carried out with relative ease.

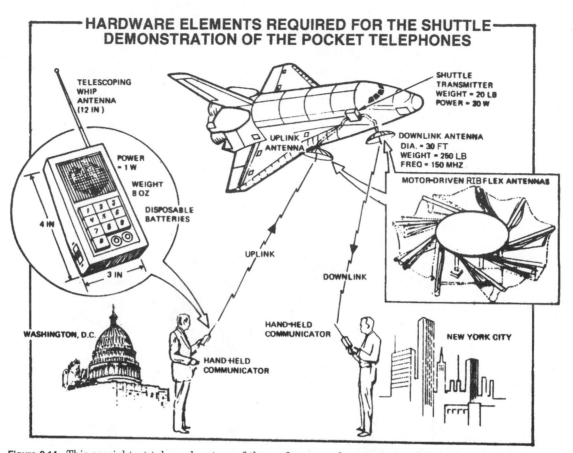

**HARDWARE ELEMENTS REQUIRED FOR THE SHUTTLE DEMONSTRATION OF THE POCKET TELEPHONES**

TELESCOPING WHIP ANTENNA (12 IN)

POWER = 1 W

WEIGHT 8 OZ

DISPOSABLE BATTERIES

4 IN

3 IN

SHUTTLE TRANSMITTER
WEIGHT = 20 LB
POWER = 20 W

UPLINK ANTENNA

DOWNLINK ANTENNA
DIA. = 30 FT
WEIGHT = 250 LB
FREQ = 150 MHZ

MOTOR-DRIVEN RIB FLEX ANTENNAS

UPLINK

DOWNLINK

WASHINGTON, D.C.

HAND-HELD COMMUNICATOR

HAND-HELD COMMUNICATOR

NEW YORK CITY

**Figure 2.14** This special test takes advantage of the performance characteristics of the space shuttle to demonstrate the capabilities of shirt-pocket telephones to relay voice messages through space. The shuttle carries two 30-ft unfurlable antennas together with a 150-MHz radio transceiver. The two pocket telephones weigh 8 oz each and produce 1 watt of RF output electrical power. For maximum impact, the study managers envisioned a brief pocket telephone conversation between the mayor of New York and the President of the United States.

whip antennas served by three large duplex antennas onboard the orbital antenna farm. Each spacecraft antenna produces 100 pencil-beams each of which covers a small region of the continental United States. Notice that all four of the communication links operate at frequencies between 4.0 and 4.6 GHz.

During the course of our engineering studies on the orbital antenna farm, our engineering team set up the plans for a low-altitude mission designed to demonstrate the capabilities of the pocket telephone portion of the overall system. Two 30-ft unfurlable antennas were to be deployed from the shuttle cargo bay as the shuttle swept over America's east coast. One version of the demonstration (Fig. 2.14) called for the president of the United States to carry on a televised pocket telephone conversation with the mayor of New York.

The demonstration we envisioned nearly 20 years ago provides a convenient introduction to later chapters in this book which deal primarily with mobile communication systems supported by orbiting satellites. Most of the systems we will review will also support shirt-pocket telephones.

## References

1. Jonathan Weber, "Stop the Presses: Papers Enter a Brave New World," *Los Angeles Times,* January 17, 1994, p. A-1.
2. Tom Logsdon, *Space, Inc.,* Crown, New York, 1988.

# Part 2

# Spaceborne Mobile Communication Systems Now in Use

# 3

# Mobile Communications on the High Seas

*Radio has no future.*

LORD KELVIN
*President of the British Royal Society, 1904*

Despite the recent growth of high-capacity optical fibers, more than 60 percent of today's overseas telephone conversations are still flowing through orbiting satellites. In fact, according to Burton Edelson and his colleagues at the International Technology Research Institute, spaceborne communication systems are "the largest and most successful of all commercial space enterprises. Such enterprises are... currently a $15 billion-per-year business which could grow to $30 billion-per-year within a decade."

Communication services linking the oceangoing vessels operated by large shipping concerns are one of the largest and fastest growing segments of the spaceborne communication market. Inmarsat, an international communication consortium based in London, England, has skillfully exploited that maritime market. Equipped with a constellation of only four geosynchronous communication satellites, Inmarsat's engineers are providing highly profitable communication hookups for oceangoing vessels plying trade routes on all the seven seas.

More than 150,000 boats weighing 25 tons or more now search out commercial fish—mackerel, tuna, shrimp, anchovies, squid. Half of them operate on voyages lasting at least several days. The private yacht market is also large—and growing. About 250,000 yachts at least 40 ft long could easily be outfitted with hardware capable of accessing the signals from today's Inmarsat satellites. Small wonder Inmarsat's alert squadron of marketeers foresee 100,000 users by 1995, with as many as a million more coming online a decade later!

A million Inmarsat users? How can they hope to find new customers in such abundant numbers? Inmarsat's marketing specialists have a convincing answer: tomorrow's growing legions of land-mobile users coupled with the maritime customers they can easily capture quickly add up to their projected total.

Inmarsat's communication links are capable and sophisticated, but even a simple system like the French Argos has captured more than 10,000 satisfied customers over the past several years. They have done it with crude spaceborne hardware capable of relaying only the simplest "telegram" messages and estimating the user's position with a 20-year-old Doppler shift approach.

## The French Argos Bent-Pipe Messaging System

The French Argos is an extremely popular and relatively inexpensive space-based communication/radionavigation system. American space vehicles, such as the National Oceanographic and Atmospheric Administration (NOAA) weather satellites, relay short telegram messages from the Argos transmitters mounted on various maritime platforms, including oceangoing vessels and drifting buoys. The Argos system employs bent-pipe relays whereby upcoming electromagnetic signals follow a sharp angular route from the buoy up to an orbiting satellite and then back down to a special computer processing facility located on the ground—where the messages are decoded and the navigation solutions are computed.

Three-dimensional navigation solutions, like the ones needed for weather balloons, are typically in error by approximately 4000 ft. Simpler two-dimensional solutions, such as the ones used in connection with oceangoing vessels, typically provide 1200-ft navigation errors. The French Argos is used for a variety of practical applications, including earthquake fault monitoring and the determination of the migration routes of relatively large animals such as caribou and moose. The Argos system is also used to monitor the status of big pipeline valves and to track hazardous icebergs ambling across North Atlantic shipping routes.

### Message Relay Services

Digital pulse trains superimposed on the carrier waves transmitted up to the NOAA weather satellites can be used to relay short "telegram" messages between Argos users at widely separated sites. These telegrams, which consist of a limited number of alphanumeric characters, typically employ short pre-arranged codes to convey quite a bit of useful information. Here are some sample messages of the type commonly broadcast by one of the trawlers in a big fishing fleet:

"ABL 15.3": 15.3 baskets of albacore tuna caught

"STJ 317": Arriving at St. Johns on March 17th

"CG": Changing main gear

Some of the digital messages relayed by the Argos system are processed onboard the NOAA satellites and rebroadcast immediately in real time to users on the ground. Other, less urgent, messages employ store-and-forward processing techniques in which the digital pulse trains are recorded onboard the satellite and later "dumped" to the ground. Message dumping occurs when the satellite moves to within line-of-sight range of one of its three ground station antennas located at Fairbanks, Alaska, Wallops Island, Virginia, and at Lannion in France.

### Bent-Pipe Radionavigation Techniques

The sketch in the upper left-hand corner of Fig. 3.1 shows how the orbital trajectories of the two NOAA weather satellites carry them near the North and South Poles in low-altitude sun-synchronous orbits. Digital messages stream up from numerous transmitters located on or near the ground to the two satellites which relay them to the ground processing center on the right. The processing center demodulates each digital signal and, using Doppler shift measurements of its carrier waves, computes the approximate geographic position of the transmitter when it was broadcasting its signals up into space.

The plaintive whistle emitted by a locomotive can be used to illustrate how the Doppler shift from the Argos navigation system can be used to determine the transmitter's position. Suppose you are standing right next to a railroad

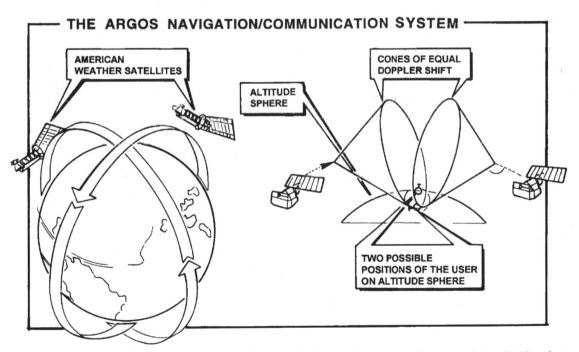

**Figure 3.1**   A French Argos transmitter on or near the ground relays an electromagnetic wave modulated with a short "telegram" message through one of America's NOAA weather satellites. The satellite, in turn, transmits the message to a special ground-based processing center which distributes it to the intended recipient. Computers at the processing site automatically determine the user's position by measuring the Doppler shift distortion of its continuous tone created by the smooth motion of the satellite as it sweeps across the sky high above the transmitter site.

track when a high-speed train passes by. As the train approaches, the waves from its whistle will be compressed to produce a higher pitch. When the train passes by and begins to recede, the waves will be stretched out to produce a lower pitch than would otherwise be observed. If you construct a rectilinear graph of frequency versus time, it will turn out to be a "step function" provided you are standing immediately beside the track.

But, if you move 3 nautical mi back away from the track and construct a similar graph of Doppler shift versus time, it will turn out to be a gentle S-shaped curve. The gentle curvature of the resulting graph arises from the fact that the systematic shift in pitch is created by the component of velocity along your instantaneous line-of-sight vector to the train. The exact shape of your Doppler-shift curve can provide a direct estimate of how far you are away from the track. Moreover, if the train has a published schedule, and it broadcasts the exact time (modulated on the carrier waves coming from its whistle), this timing information can be used to pinpoint your lateral location along the track.

The Argos navigation system employs conceptually similar, but somewhat more complicated, position-fixing techniques. Of course, Argos works with electromagnetic waves rather than sound waves. The ground-based processing center measures the shape of the Doppler shift curve as the satellite travels along its orbit from horizon to horizon. Then, in essence, the computers at the processing center execute a curve-fitting routine to determine the shape of the Doppler shift curve—which indicates how far the ground-based transmitter is from the satellite's ground trace.

As the sketch on the right in Fig. 3.1 indicates, an ambiguity exists in the Argos navigation solution because the transmitter could be 100 nautical mi to the *east* or 100 nautical mi to the *west* of the satellite's ground trace and still produce a Doppler shift trace of exactly the same shape. But, fortunately, the earth's rotation distorts the Doppler shift trace just enough to eliminate the solution ambiguity.

## Sun-Synchronous Orbits

The NOAA weather satellites employed by the Argos system are launched into 435-nautical mi sun-synchronous orbits with orbital inclination of 98.5 degrees. At that altitude, each NOAA satellite completes exactly 14 orbits per day. Each circuit around the earth takes about 102 minutes.

As the satellite travels around its orbit, any users within 2600 nautical mi of its ground trace can access it to relay their digital pulse trains and their carrier waves to the ground-based processing center. Users on the ground have access to the two satellites 7 to 28 times per day, depending on their latitude location. Those who are near the North and the South Poles have the largest number of opportunities to access the satellites. Those near the equator have the smallest number of satellite-access opportunities.

The geometrical characteristics of a sun-synchronous orbit are sketched in Fig. 3.2. A sun-synchronous satellite is placed in a near-polar orbit with precisely the right altitude and inclination to make its orbit plane twist approxi-

## USING WEATHER SATELLITES TO SAVE LIVES

If a special weather satellite had not been available when Serge Goriely's four-wheel-drive Citroen whipped out of control in a remote African desert, he might not be alive today. Goriely, a 21-year-old professional racecar driver, suffered a fractured skull and lay motionless beside his crushed vehicle after it crashed, rolled over several times, and threw him out unconscious into the tightly packed desert sand.

Fortunately, his car was equipped with an experimental search-and-rescue beacon that was automatically activated when his car went off the road. Immediately, it began sending a distress signal into space that was relayed back to Paris, where it was picked up only 17 minutes later. A medical specialist was promptly flown to the scene of the accident, arriving there 79 minutes after the crash. He patched Goriely back together, then admitted him to a nearby hospital for several days of recuperation before his colleagues knew for sure that he would live long enough to join the 344 others whose lives had been saved by a search-and-rescue satellite called SARSAT.

Teams of technicians in the United States, Canada, France, the former Soviet Union, and seven other participating countries work together to make sure SARSAT stays on the air. Emergency beacons—space-age cries for help—stream up to the satellites from planes, boats, and even battered racecars for immediate retransmission by American and Russian satellites. Relay stations on the ground then pass the information on to rescue centers assigned to dispatch appropriate rescue forces. The satellites are designed to relay specially coded messages that tell who (or what) is in trouble and the approximate location of the distress beacon.

Before the SARSAT system was available, average notification time for a missing aircraft was 36 to 48 hours. However, studies show that, if lives are to be saved, rescue must usually be accomplished within 24 hours. With four active SARSATs in the constellation, an emergency signal from Africa, or any other remote location, can always be picked up by ground monitoring systems within 1 hour.

mately 1 degree per day (360 degrees per year). Thus the angle between its orbit plane and the radius vector to the sun remains nearly constant at all times throughout the entire year.

As the sketch in Fig. 3.2 indicates, the orbit plane of a satellite twists because the gravity created by the earth's equatorial bulge pulls on the satellite in a systematic manner. When the satellite is in the Northern Hemisphere, the equatorial bulge pulls it *down* toward the equatorial plane; when it is in the Southern Hemisphere, the equatorial bulge pulls it *up* toward the equatorial plane. These systematic gravitational perturbations create a gyroscopic phenomenon causing the orbit plane to twist in space, typically a few degrees per day.

A satellite in a sun-synchronous orbit with repeating ground trace geometry achieves similar lighting conditions every time it arrives over the same spot on the earth. Constant lighting angles help make the NOAA satellite's weather observations more standard, useful, and easier to interpret. The Argos does not benefit from the satellite's unique orbital geometry; it is merely a piggyback payload along for the ride.

Argos is used for maritime communication and positioning services. It also relays messages and positioning information from unmanned weather balloons, maritime buoys, seismic stations, pipeline valves, and from large migrating animals. Argos has also, on occasion, been used to track Arctic and Antarctic icebergs floating in the frigid sea.

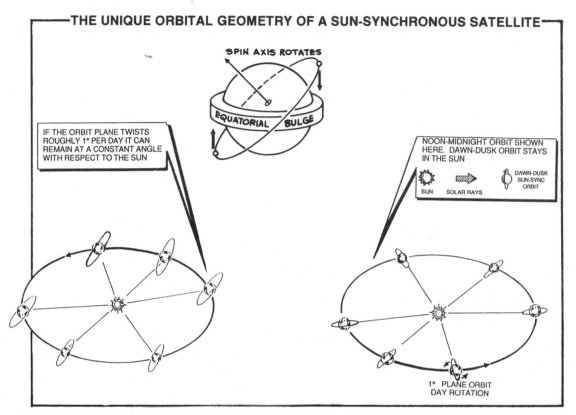

**THE UNIQUE ORBITAL GEOMETRY OF A SUN-SYNCHRONOUS SATELLITE**

SPIN AXIS ROTATES

EQUATORIAL BULGE

IF THE ORBIT PLANE TWISTS ROUGHLY 1° PER DAY IT CAN REMAIN AT A CONSTANT ANGLE WITH RESPECT TO THE SUN

NOON-MIDNIGHT ORBIT SHOWN HERE. DAWN-DUSK ORBIT STAYS IN THE SUN

SUN    SOLAR RAYS    DAWN-DUSK SUN-SYNC ORBIT

1°  PLANE ORBIT DAY ROTATION

**Figure 3.2**   A sun-synchronous satellite is launched into a near-polar orbit with a specific altitude-inclination combination. Gravitational perturbations induced by the earth's equatorial bulge cause the satellite's orbit plane to twist 360 degrees each year, thus compensating for the gradually changing geometry as the earth travels around the sun. One benefit of a sun-synchronous orbit is that solar illumination angles on the satellite and on the ground below remain nearly constant throughout the satellite's mission life.

## Tracking Icebergs with the Argos Satellites

In 1912, the "unsinkable" British luxury liner *Titanic* slurped its way down toward Davy Jones's locker when it suddenly collided with a large Arctic iceberg. That incident received worldwide publicity, but actually it represented only one of many tragic encounters between big chunks of hazardous sea ice and maritime vessels.

Each year 10,000 icebergs are created in the Northern Hemisphere. About 5000 of them reach the open ocean, and, on average, 300 pass below 48 degrees north latitude where they become a major hazard to North Atlantic shipping. During the iceberg season, which lasts about 5 months, shipping routes are lengthened as much as 30 percent so large vessels can avoid the worst concentrations of floating ice. Gangling oil platforms in northern waters are also at hazard, and, on occasion, they must be abandoned when icebergs move into their vicinity.

Of course, no region in the North Atlantic is entirely free from iceberg hazards. Consequently, Arctic icebergs are carefully tracked by the International Ice Patrol, an organization that is managed by the U.S. Coast Guard, but financed by various maritime nations operating in the North Atlantic. The International Ice Patrol sends ships and airplanes into the North Atlantic to locate as many icebergs as possible. Targets of opportunity are also reported by commercial and military vessels. Six radio stations broadcast iceberg warnings twice per day so ships in the area can safely adjust their movements.

So far, the Argos system has been used to track at least a dozen icebergs bobbing along in the chilly waters of the North Atlantic. In one test (Fig. 3.3), an 80-lb Argos transmitter was parachuted onto the surface of a flat Arctic iceberg. The transmitter was then used to relay crude position coordinates to National Aeronautics and Space Administration (NASA) engineers at Wallops Island, Virginia. Over a 3-month interval the iceberg traveled along an ambling, erratic trajectory across 200 mi of open ocean. On the average, it moved less than 5 mi/day, well below walking speed. About 20 percent of the time, the iceberg was essentially stationary because it was either permanently or intermittently grounded.

---

## TRACKING ARCTIC ICEBERGS WITH ORBITING SATELLITES

In 1912, the "unsinkable" British luxury liner *Titanic* sank when it collided with an Arctic iceberg. That particular iceberg collision is well known to the general public, but at least a dozen other, less famous, collisions between maritime vessels and iceberg fragments have occurred in this century. During the 5-month iceberg season, some shipping routes must be lengthened as much as 30 percent to skirt the worst concentrations of floating ice. This ties up valuable cargoes and cuts the productivity of shipborne crews.

To minimize the probability of damaging collisions, the International Ice Patrol, which is managed by the U.S. Coast Guard, gathers and disseminates daily information on iceberg sightings. Radio stations manned by Ice Patrol personnel broadcast warnings to ships in the North Atlantic twice each day.

Researchers have attempted to destroy icebergs by bombing, torpedoing, shelling, ramming Arctic icebergs, and by painting their surface with lampblack or charcoal to increase melting rates. In one test, twenty 1000-lb bombs were dropped on a 250,000-ton iceberg, but only 20 percent of it was chipped away. Of course, icebergs are often much larger than 250,000 tons. A few of them are 50 mi long.

Because of their high degree of accuracy and their continuous availability, the Navstar satellites could provide an attractive method for tracking the 300 or so icebergs that pass below the 48th parallel in an average year. A Navstar receiver would be dropped onto the iceberg to broadcast its current position to any nearby vessel. Similar, but less capable, spaceborne tracking techniques have already been tested at least a dozen times. In one such test, researchers from Chevron parachuted an 80-lb Argos transmitter onto the surface of an Arctic iceberg. It then relayed crude position coordinates to a NASA tracking station through American weather satellites.

Over a 3-month interval, the iceberg ambled across 200 mi of open ocean. On the average, it traveled less than 5 mi/day, well below walking speed. About 20 percent of the time, it was stuck on the bottom or jammed against various islands. Unlike the Argos method of tracking icebergs with its time-consuming data relays and remote computer processing techniques, the Navstar approach could provide precise tracking with real-time warnings for any ships that might be at hazard from Arctic ice.

## ICEBERG TRACKING WITH THE ARGOS NAVIGATION SYSTEM

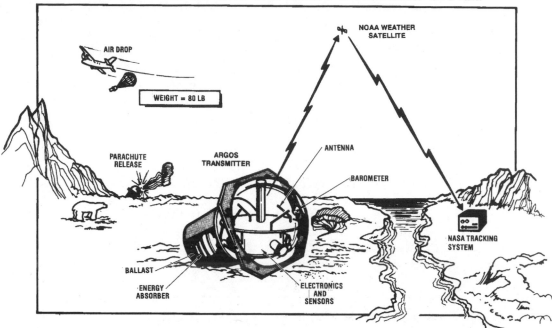

**Figure 3.3**   So far, American oceanographers have parachuted Argos radio transmitters onto the surface of at least a dozen Arctic icebergs. Once in place, the transmitter relays radio signals through NOAA weather satellites to provide Doppler shift measurement that can be used to reconstruct the convoluted trajectory of the iceberg to an accuracy of about 1200 ft. With tracking by the Navstar satellites, accuracies can be improved and real-time warnings can be provided to any ships sailing into the vicinity of a floating iceberg. This will enhance the safety of the crews and allow vessels plying North Atlantic trade routes to reach their destinations sooner, more safely, and with less consumption of valuable fuel.

The Argos system turned out to be a useful tool for tracking icebergs experimentally, but the positioning measurements it provided were not very accurate and they were not obtained in real time. Consequently, better methods are required to keep track of the 300 icebergs that pass below the 48th parallel in an average year. More precise tracking with the Navstar Global Positioning System (GPS) is a definite possibility. The Navstar radionavigation system can provide continuous, real-time monitoring of the hazardous icebergs that menace North Atlantic shipping lanes. The key features of the Navstar navigation system are briefly discussed later in this chapter.

## Inmarsat's Maritime Communication Satellites

With a constellation of only four geosynchronous communication satellites, Inmarsat engineers are able to relay two-way voice messages and computer data to and from remote installations, most of which are mounted on relatively large oceangoing vessels. Once the user selects the proper satellite, the appropriate telephone number is dialed in the manner of making an ordinary international telephone call. Communication frequencies are in the L-band portion of the frequency spectrum in the vicinity of 1600 MHz.

As Fig. 3.4 indicates, the four Inmarsat satellites are positioned along the equator over South America, in the mid-Atlantic, over the Indian Ocean, and above the Pacific about 30 degrees off the east coast of New Guinea.

Between 25,000 and 30,000 customers regularly access Inmarsat's mobile communication links. Most of their communication facilities are today permanently installed onboard relatively big ships. Inmarsat officials are anticipat-

## THE INMARSAT PORTABLE SATELLITE TELEPHONE SYSTEM

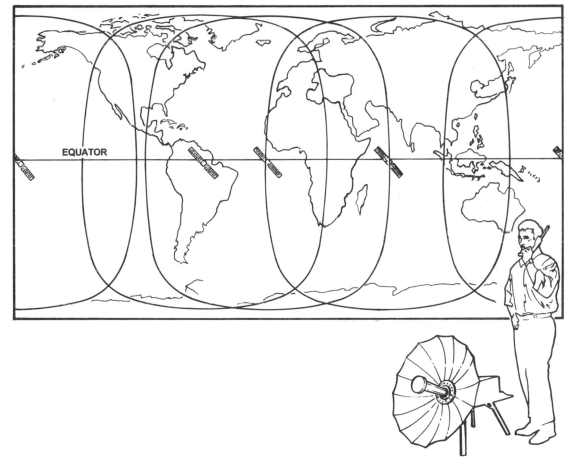

**Figure 3.4** With a family of four geosynchronous communication satellites, Inmarsat provides personal communication services for about 30,000 maritime users. Except for two small regions surrounding the North and South Poles, their constellation provides continuous coverage for the entire surface of the globe. Most of the Inmarsat installations are permanently mounted aboard fairly large oceangoing vessels. But, as the sketch at the bottom of this figure indicates, portable land-mobile terminals are available that can be packaged in a suitcase-sized container and carried from place to place aboard commercial jetliners. (*Photograph courtesy of Comsat Corporation.*)

ing spectacular growth in their business—100,000 users by 1995, a million by 2005! The first step in this new revolution in personal communications will come with Inmarsat 3, two copies of which are slated to ride into space aboard Atlas-Centaur boosters. Another Inmarsat 3 will be lofted by a European Ariane, and a fourth will sit astride a four-stage Russian Proton booster.

Inmarsat offers various levels of service at different rates for its mobile phone customers. Their oldest service, Inmarsat-A, was designed to accommodate top-of-the-line customers who require a variety of high-quality communication services and who can afford to pay premium rates. Inmarsat-A provides analog FM data and voice modulations at $6 to $10 per minute. The smallest shipboard Inmarsat-A installations weigh about 80 lb each. Inmarsat-A's data rate is 9600 bits/s.

Inmarsat-A will be supported for at least another decade, but it is gradually being replaced by Inmarsat-B, which provides digital voice modulations for about $7 per minute (with off-peak hour discounts).

The newest Inmarsat service is Inmarsat-M, which sacrifices a bit of voice quality and data rate to achieve dramatic reductions in weight, size, and cost of equipment. Calls on Inmarsat-M cost about $5.50 per minute. The data rate for Inmarsat-M is only 2400 bits/s, so users looking for high-speed data or compressed video will not likely be very happy with Inmarsat-M, which does not support telex at all. Briefcase-sized Inmarsat-M terminals weighing only 30 lb each are on sale for $15,000. This compares favorably with the $30,000 to $40,000 price tags on old-style Inmarsat-A shipboard terminals, some of which weighed as much as 300 lb.

Inmarsat also provides a store-and-forward data-only service called Inmarsat-C. The shipboard equipment for Inmarsat-C costs only about $4500 and messaging services are priced at $1.12 for 1000 bits (about 125 alphanumeric characters or 20 English words). Further cost reductions for voice and data services are anticipated when the Inmarsat 3 satellites reach their assigned locations in space. The small electromagnetic spot beams illuminating the ground from Inmarsat 3 will greatly increase communication capacity to further drive down costs.

## Portable Equipment Modules

During the Persian Gulf War, international press correspondents stationed in Baghdad managed to maintain regular telephone contact with the outside world using portable Inmarsat voice terminals. Magnaphone units of this type, which are manufactured by the Magnavox Corporation, weigh only 50 lb. They are, in fact, so light and compact that they can travel on commercial jetliners as ordinary airline luggage! Some models employ rigid parabolic antennas, but unfurlable antennas, such as the one sketched in Fig. 3.4, are also available in the commercial marketplace.

Glocam, Inc. of Rockville, Maryland makes a unit called the 2500t which is roughly the size of a standard briefcase. To deploy its "antenna," the user merely opens the lid of the briefcase and manually adjusts its position until the built-in signal-strength meter displays its maximum value. Batteries power the unit for a maximum 1-hour conversation. About 20 companies worldwide are producing Inmarsat-M transceivers of various types.

## Norway's Clever Scheme for Expanding Inmarsat's Coverage Regime

Inmarsat's geosynchronous communication satellites normally provide communication coverage for a broad band around the earth's equator ranging from about 75 degrees north to 75 degrees south latitude. The satellites are actually above the horizon for latitudes closer to the poles, but, unfortunately, they tend to be very near the horizon at these latitude extremes so stray reflections from the sea tend to create destructive interference.

However, clever Norwegian engineers have found a way around this apparent limit, so their fishing boats are now able to maintain contact through Inmarsat satellites at much higher latitudes. Some Norwegian vessels are now maintaining contact even when they venture up to 82 degrees north latitude to reach the rich fishing grounds around the Island of Spitzbergen. In a series of high-latitude tests, Norway's engineers noticed an interesting phenomenon that helped them solve the coverage problem. From sea level upward, there were separate layers of undistorted signals alternating with layers of noisy, unusable signals, not unlike the layers of a wedding cake. The engineer's approach toward latching on to the usable signals was simple but elegant. They merely installed *two* antennas on each fishing vessel positioned at different heights with an automatic switch rigged to select the best of the two signals. This clever technique works like a charm! Now their fishing vessels can get good coverage all the way up to 82 degrees latitude and slightly beyond.

## Services Provided by the Inmarsat Satellites

In 1992, when the crew of the *Rainbow Warrior,* a special Greenpeace ship, attempted to halt nuclear testing on Mururoa Atoll in the Pacific, they realized that their efforts would have far more impact if they could gain worldwide television coverage. That is why Greenpeace planners carried video cameras to the demonstration and rigged their ship with top-of-the-line Inmarsat-A equipment to link it with their London headquarters. Their link-up was accomplished by using a store-and-forward data transmission system called Skylink, coupled with computerized data compression techniques.

The Greenpeace master plan called for their representatives to breach Mururoa's 12-mi exclusion limit, race on shore without being detected, and take coral samples for later analysis. Unfortunately, they managed to get only half that far. About 6 mi from shore their boat was intercepted by a contingent of marines who swiftly escorted the entire Greenpeace crew out of the area. However, video images of the distant shoreline and the foreground altercation were successfully transmitted back to London with video images appearing on the air less than 1 hour later.

The Greenpeace vessel was rigged with $80,000 worth of Inmarsat communication equipment, and renting the 56-kbit/s Inmarsat-A data link cost the organization another $300,000. But mission planners maintained that the result was well worth the cost, in the view of the worldwide publicity their protest received.

The equipment carried onboard the *Rainbow Warrior* included a standard Inmarsat-A shipboard terminal with antenna array, a video recorder, color monitor, video codec image compression software, and a Toshiba laptop computer programmed to handle the necessary store-and-forward images. Even at Inmarsat-A's high-speed 56-kbit/s data transmission rate, a 14-min transmission interval was required for each minute of video transmitted to the London headquarters. On the London end another codec software routine converted the data files back into the original video, which was played over the air at normal speed.

Of course, most practical applications of Inmarsat communication hardware center around much more mundane shipboard purposes such as maintenance and repair. Today's ships operate with the smallest possible crews. Most crew members know how to operate the equipment, but many do not know how to keep it in repair. That is why Inmarsat's remote video can be so useful in today's highly competitive world. Emergency medical treatment may also depend on static-free and reliable ship-to-shore communication links.

Recently some dramatic results were reported in conjunction with a demonstration of space-age video communication techniques as reported in an article in *Ocean Voice*. During that demonstration, a maritime maintenance specialist sat in front of a video monitor inside a crowded exhibit hall in Germany. A telephone set in his hand linked him by audio to the ship. He was also aided by electronic outputs from an Apple Macintosh computer.

For purposes of the demonstration the maintenance specialist had been assigned to inspect the engine room of a distant vessel for any problems, with emphasis on a faulty nozzle the crew had been fretting about. They needed to know, in particular, whether to order spare parts, put in for emergency repairs, or continue the voyage without concern.

"Switch now to the camera in the engine room, please," the maintenance specialist requested over the telephone handset. "I would like to check some of the systems there."

A second or two later, the engine room came into view. "Can you zoom in on the meters in the center of the picture, please." The shipboard camera operator responded by zooming in and focusing on two battered gauges with jiggling needles.

"I can see the gauges very clearly now; the pressure seems OK," the engineer said in reassuring tones. "Now I would like to have a look at the centrifugal filter to see if the connecting ring is correctly attached and there is no leakage."

A few more camera adjustments followed by more verbal exchanges brought this final evaluation: "I can see there is a small leakage, but that's OK, it's correctly attached. It can wait until the next routine maintenance period."

With other, similar audio and video systems, shipboard personnel can stay in contact with experts all over the world, while teams of technicians on both ends see exactly the same computer displays. Whenever anyone on either end points a cursor, adds data, or makes an electronic sketch, both screens change in instant response.

With today's smaller crews and increasingly sophisticated shipboard systems, maritime experts onboard the ship may know how to operate the equipment, but not how to keep it in repair. Fortunately, today's highly capable mobile communication systems are beginning to allow shore-based experts to come on board at any time by electronic means.

## Computerized Ship-Routing Techniques

"Only a man who has commanded a ship at sea can know the burden of responsibility borne by a modern ship's master." So states a colorful pamphlet published by Oceanroutes, a space-age enterprise that helps large commercial vessels find their way through dangerous weather. "For he knows that despite the clear calm of any moment on the sea, there lurks ahead the threatening storm, the pall of fog, the edge of ice.... Many morning watches will find him huddled behind a wind screen after a night of staring into the foggy murk. Because always the responsibility for crew, and cargo, and ship is his."

The master of a ship does, indeed, have worrisome responsibilities, but those responsibilities are getting easier now that the experts at Oceanroutes are available to lighten the load by using their computers to process and combine information from Inmarsat communication satellites, international weather satellites, and the Navstar GPS. For a fee averaging $700 per voyage, they provide efficient real-time routing information to help harassed ship captains find the safest, most efficient trajectories across the sea.

After the Oceanroutes experts gather the necessary data, they process it in real time using high-speed computers, then transmit the results through Inmarsat communication satellites twice a day to oceangoing vessels. From their comfortable offices in Palo Alto, California, and at several other locations around the globe, those diligent specialists work with a thousand ships in a routine month. Each recommended route is worked out for that particular ship "on that specific voyage, with the given cargo load, status of trim and draft, with the ship's own distinctive speed and sea-handling characteristics."

The computer program they use emphasizes emerging weather, but it also takes into account currents and fog, choke points and navigational hazards, and sea ice in northern regions. Some cargos, such as oil and fruit, are temperature-sensitive; others, such as automobiles and heavy machinery, may shift under heavy waves. Still others have time-critical deliveries. The program successfully takes these and other factors into account whenever it makes routing recommendations.

According to Alan Cima at Oceanroutes Headquarters in Palo Alto, California, a typical shipping concern saves $30 to $40 for every $1 spent on data from Ocean-

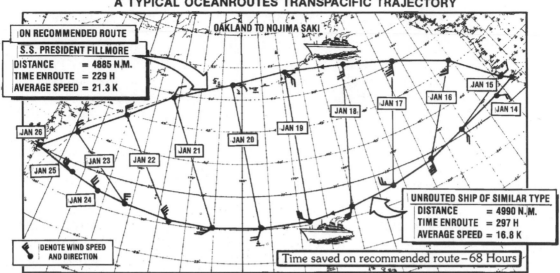

**Figure 3.5**  The experts at Oceanroutes, Inc. in Palo Alto, California use high-speed computers to develop safe and efficient shipping trajectories like the one at the top of this figure. Every routing computation takes into account real-time weather conditions, the physical characteristics of the ship, and the wishes of the ship's master—who receives an updated trajectory twice each day. Inmarsat satellites provide most of the communication links, whereas the satellites in the Navstar constellation provide much of the positioning information. The cost of the service for a typical voyage is $700, a fee that is repaid 30 to 40 times over by shortened travel times and more efficient maritime operations.

routes. A reference trajectory generated by the Oceanroutes computer is presented in Fig. 3.5 together with a similar trajectory mapped out without the aid of computers. Notice that the computer-generated real-time route shaves off 68 hours in total travel time.

Updates are provided every 12 hours thanks to Inmarsat's constellation of geosynchronous communication satellites. If hazardous weather threatens to close in anywhere along the way, the ship's master is alerted and given an instant recommendation for a new course to keep crew and vessel out of harm's way. If rerouting is impossible, the captain is notified of the force and duration of hazardous winds and waves. So far in more than 43,000 crossings aided by Oceanroutes information, travel times have been reduced an average of 4 hours in the Atlantic, 8 in the Pacific. Operating a large oceangoing vessel costs as much as $1000 an hour, so time savings alone translate into substantial reductions in cost. In addition, other expenses are also routinely reduced.

In the early 1960s when Oceanroutes services were not yet available, the cost for repairs of weather-damaged ships ran from $32,000 to $53,000 in an average year. Today these costs have dropped to only about $6000. Cargo damage has also decreased. One international auto dealer told Oceanroutes researchers that his damage claims for cargo had dropped over $500,000 per year. Another big Oceanroutes customer discovered that damage claims were running less than half those in the industry as a whole.

## The Navstar GPS Radionavigation System

As Fig. 3.6 indicates, the Navstar GPS satellites orbit the earth in 10,898-nautical mi orbits in six orbit planes, each tipped 55 degrees with respect to the earth's equatorial plane. The complete constellation consists of 21 Navstar satellites, plus 3 active on-orbit spares. Each Navstar GPS satellite (Fig. 3.7) weighs about 2000 lb and is constructed from 65,000 separate parts. Winglike solar arrays on the sides of each satellite generate 710 watts of electrical power

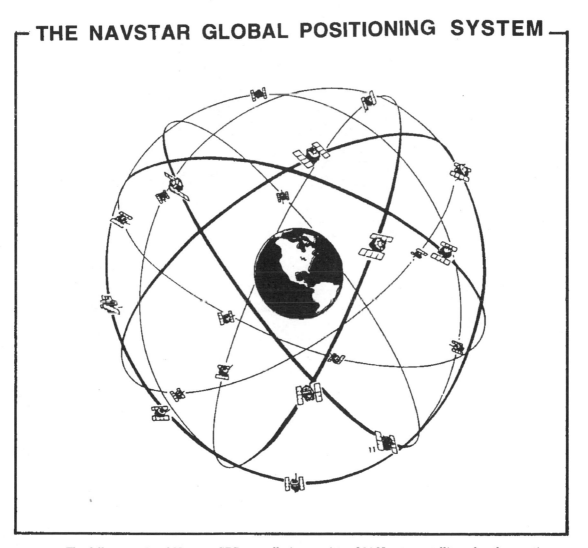

**Figure 3.6** The fully operational Navstar GPS constellation consists of 21 Navstar satellites plus three active on-orbit spares traveling around 12-hour circular orbits 10,898 nautical mi above the earth. A feedback control loop helps maintain a continuous earth-seeking orientation for the 12 navigation antennas mounted on the main body of the spacecraft. A second feedback loop helps maintain a sun-seeking orientation for the satellite's two winglike solar arrays.

# A TYPICAL NAVSTAR NAVIGATION SATELLITE

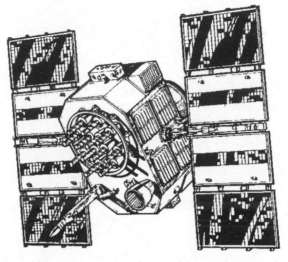

**Figure 3.7**  Each Navstar navigation satellite is constructed from 65,000 separate parts. Yet it is designed to operate for 7.5 years or 580 million mi, whichever comes first. The solar cells mounted on its winglike solar arrays generate 710 W of electricity to power the transmitter and the other satellite subsystems. Internal temperatures are maintained within narrow limits by thermostatically controlled louvers and goldized Mylar insulation blankets wrapped around some of the satellite's more temperature-sensitive components.

to drive the onboard transmitters that broadcast precisely timed L-band navigation pulses down toward users on the ground.

The satellites and their ground support equipment are being financed by the U.S. Department of Defense, but their navigation signals are available free of charge to anyone, anywhere, who cares to use them. Someday soon the system may serve millions of users on land, in the air, and at sea.

Signals from the satellites in the Navstar constellation help position many of the ships that subscribe to the services provided by Oceanroutes. Positioning measurements and routing instructions are sent through Inmarsat's communication satellites. The newer Inmarsat 3 satellites are also supplementing the services provided by the Navstar satellites in a number of important and interesting ways. Some of these service supplements are discussed in the next section.

The Navstar constellation helps mariners, commercial and amateur pilots, military personnel, and many other enthusiastic users fix their positions with extraordinary ease and precision. Like most of its competitors in the world of navigation, the Navstar GPS is a radionavigation system that uses passive triangulation to pinpoint the user's position. Positions are usually displayed in three mutually orthogonal coordinates such as longitude, latitude, and altitude. For military users, a positioning accuracy of 50 ft is usually achieved even under high-dynamic conditions. Civilian users, more typically, achieve a 330-ft accuracy in their positioning solutions.

Each Navstar satellite broadcasts a prearranged sequence of timing pulses to blanket the full disk of the earth with modulated positioning signals. When the receiver picks up a timing pulse, it multiplies the measured signal travel time by the speed of light (186,000 mi/s) to obtain the range to the satellite.

In theory, three range measurements of this type could be used to nail down the user's current location in three dimensions. In practice, however, ranging information from a fourth satellite is needed to compensate for any clock errors in the receiver. This approach allows the receivers to be equipped with inexpensive quartz-crystal oscillators rather than precisely synchronized atomic clocks. A crystal oscillator, which is smaller than the head of a two-penny nail, would require nearly 3 years to lose (or gain) a single second. Although this may seem impressively accurate by ordinary standards, the atomic clocks carried onboard the satellites are 10,000 times more stable and accurate than quartz-crystal oscillators.

Indeed, the key to the Navstar's extraordinary precision lies in the ability of the satellite's atomic clocks to keep amazingly accurate time. In a radionavigation system even the smallest timing errors are intolerable, because a radio wave travels 1 ft in a *billionth* of a second. Consequently, every billionth-of-a-second timing error creates at least a 1-ft error in navigation. To achieve the desired 50-ft accuracy on a global basis, the satellite clocks must always be mutually synchronized to within 13 billionths of a second. Navstar's designers achieved this difficult goal by constructing miniature atomic clocks that are so stable and accurate they would require 300,000 years to lose or gain a single second and by resynchronizing the satellite clocks from the ground at least once every day.

In the 1970s, when preliminary concepts for the Navstar navigation system were first being investigated, a typical atomic clock was roughly the size of a household deep freeze—far too big and heavy to fly into space. However, space-age miniaturization techniques have slimmed down today's atomic clocks to only 30 lb, or even less. Each modern Navstar satellite is equipped with four mutually redundant atomic clocks: two cesium clocks and two rubidium clocks. During the mission if one clock fails, another is switched on to take its place.

## Inmarsat's Geosynchronous Overlay

Inmarsat's engineers are installing piggyback payloads aboard their Inmarsat 3 geosynchronous satellites to provide L-band-compatible GPS-type navigation signals. The extra cost of the flight hardware capable of giving the world this valuable service is estimated to be only about 5 percent the cost of building and launching an additional GPS satellite. Inmarsat's binary pulse sequences will be generated on the ground for relay through their geosynchronous communication satellites. These new satellites will enlarge the conventional GPS constellation of 12-hour satellites to provide improved navigation accuracy and to help make the GPS coverage more robust. In addition, the piggyback payloads will also provide another bonus service. With the proper ground-based hardware, the current integrity (usability) of the GPS satellite signals can be monitored in real time. If a discrepancy is ever discovered, this information will be relayed through the Inmarsat 3 satellites, thus providing a warning to critical users not to rely on that particular GPS satellite.

## A Brief Introduction to Project 21

In order to serve the communication needs of the next generation of mobile users on land and at sea, Inmarsat officials have been sponsoring engineering studies aimed toward developing an ambitious new mobile spaceborne communication system, code-named "Project 21."

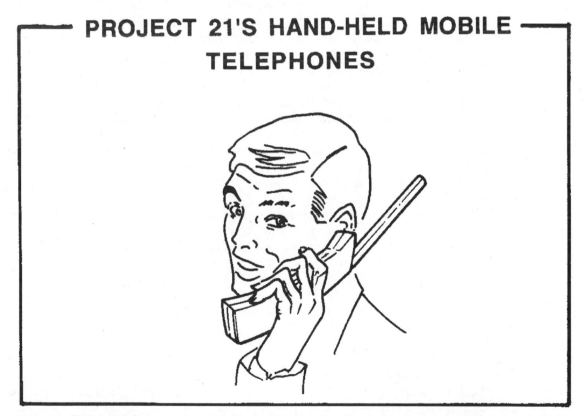

**Figure 3.8**  Project 21 is being masterminded by Inmarsat's innovative communications engineers. A constellation of Project 21 satellites positioned in medium-altitude and/or geosynchronous orbits will provide voice-messaging relay links for a worldwide class of business-related users. Until late 1993, a low-altitude constellation was in contention for the service, but after a series of preliminary studies, Inmarsat's experts abandoned that approach, primarily because of its relatively high cost.

Some of the details of Project 21, which may consist of a mix of satellites positioned at different orbital altitudes, will be discussed in Chapter 7. Target markets frequently mentioned by Inmarsat's marketeers include shipping and international business travelers, lifeboats, noncommercial aircraft, developing countries, and areas of the world where cellular services do not yet exist. A typical hand-held telephone of the type that will access the Project 21 system is presented in Fig. 3.8.

# 4

# Spaceborne Land-Mobile Communication Systems

*I think there is a world market for about five computers.*

THOMAS WATSON
*IBM executive, 1958*

Industry observers who get excited about future mobile communication satellite systems tend to focus their attention on voice, not data. But industry excitement aside, many of today's most successful land-mobile communication systems cannot handle voice at all; they transmit only short alphanumeric "telegram" messages to the intended recipients. Fortunately, most land-mobile customers are considerably more interested in efficiency than sex appeal.

Consider the trucking companies who subscribe to the OmniTRACS messaging system, for instance. Their average driver sends six or eight short alphanumeric messages each day in the form of digital pulse trains relayed through geosynchronous communication satellites. None of their messages seem likely to pull in literary prizes for content or readability, but they are chock full of useful information to help dispatchers manage big fleets of trucks with unprecedented efficiency.

Each digital message answers one or more important questions: Where is that South Dakota 18-wheeler? Is its refrigeration equipment keeping the chickens fresh? Is the truck's transmission operating at peak efficiency? Or does it need prompt repairs? Has the driver been barreling down the interstate? Or is he gradually slipping behind schedule?

Armed with timely and reliable answers to these and a number of other similar questions, fleet management becomes an exact science that can be conducted almost as efficiently as if the drivers and their rigs were all in the same room with their fleet management personnel and their high-speed digital computers.

Digital pulse trains streaming through orbiting satellites are also being used in nationwide paging services and in the tracking of hazardous shipments and high-value cargo containers.

## The Critical Importance of Digital Data Relay

*Satellite Communications* reporters Janet Dewar and Martha Cooley are totally convinced that spaceborne digital data relay systems will, in future years, have revolutionary impacts on ordinary private citizens. "The transmission of digital data might best be called a watershed technology," they say, " on the order of the wheel, the steam engine, and electricity in terms of its profound effects on human experience."

Dewar and Cooley go on to explain how digital processing techniques are beginning to blur the distinction between computers and telecommunications while, at the same time, "heightening computing power, improving quality, and lowering costs." Moreover, they note, digital modulation techniques "are helping to foster the ability to compress large amounts of information into less space for transmission purposes."

Digital implementations are easy to spot in ordinary consumer products, including such obvious examples as digital wristwatches, compact disks, and digital bathroom scales. Digital technology is also making it possible for today's experts to build efficient land-mobile communication systems based in space.

The most successful systems now operating along the space frontier transmit information in the form of binary pulses rather than voice. The amount of data they can transmit is strictly limited, but the approaches they are pioneering

---

### THE FIRST LARGE-SCALE DIGITAL COMMUNICATION NETWORKS

Contrary to popular belief, the first large-scale digital communication systems did not arise with the invention of the telegraph. Nationwide messaging networks, in fact, preceded the telegraph by a hundred years. They were in operation during the reign of Napoleon Bonaparte, who helped fund their construction.

The first such systems were built in the 1770s by the French priest Claude Chappe and Niclas Edelcrantz, a Swedish nobleman. Their innovative method of long-distance communication employed a series of tall towers extending along irregular lines between the major cities sitting on the French countryside. At the top of each tower stood a signalman equipped with a large signaling device and a telescope so he could pick up and retransmit coded messages coming in from other nearby towers. A codebook filled with simple numbered messages rounded out his collection of equipment.

In different regions and during different historical eras, the tower-mounted signaling devices were built in several variations. One early version employed a large pendulum clock rigged with one big hand that made one complete revolution every 30 seconds. Mounted beside the face of the clock was a flat panel painted black on one side, white on the other. As the hand of the clock passed through each of 10 angular positions, the signalman flipped the flat panel at the crucial moment to black or white, thus transmitting a single decimal digit. Various strings of digits corresponded to specific prearranged messages printed in the book.

An alternate signaling approach used a mechanical semaphore similar to the "outstretched arms" used even today by maritime nations. Still another employed counterweighted mechanisms that opened and closed shutters under human control.

In 1804, Napoleon Bonaparte ordered the construction of a long row of signaling stations linking Paris with Venice, 450 mi away. A year later, with generous infusions of military funds, the network was enlarged to cover nearly every major city throughout France. Four branches radiated from Paris, for instance, toward the four points of the compass, often paralleling the old stagecoach lines.

are inevitably paving the way for tomorrow's far more capable digital voice systems that are slated to employ *Star Trek*-style hand-held personal communicators.

Bruce D. Nordwall at *Aviation Week*'s Washington bureau has been tracking the many advancements in tetherless communications for the past dozen years. He, too, is convinced that digital transmission techniques have revolutionary potentialities. "The convergence of wireless communication and computer technologies promises a rich new consumer market in mobile communications," he observes. "Instead of being desk-bound by personal computers and faxes that communicate over telephone lines, people will be able to carry personal communication devices that use satellite or cellular links anytime, anywhere."

America's observant space-age marketeers are already beginning to notice the first convincing bits of evidence that Bruce Nordwall is correct in his assessment of the profits to be made in digital transmission techniques. Success stories abound in the trucking industry, for instance, and in nationwide spaceborne paging services. In both cases digital telegram-type messages are being transmitted over the airwaves. But as Nordwall and his colleagues see it, digital relay will grow and prosper in parallel with the more appealing voice services well into the next century, and beyond.

"Although most of the interest so far has been in voice, the biggest growth will occur in data communication," notes Ira Brodski, president of Datacomm Research Co. "The impact of mobile data will be as great, if not greater, than the personal computer. It will define new standards for personal interaction and timely access to information."

Dr. Gerard O'Neill, a highly successful physics professor at Princeton University, was a pioneer in a digital messaging technique that can link large numbers of land-mobile users. Dr. O'Neill participated in Rockwell International's space industrialization studies described in the previous chapter. He came to our Seal Beach, California facility several times, especially when we sponsored meetings between O'Neill and the other four consultants we hired to help with some of the research.

## Geostar's Geosynchronous Messaging Services

The Geostar Corporation was an early front-runner among companies offering satellite-based digital communication services to ordinary private citizens. Geostar officials managed to raise more than $49 million from private investors, venture capitalists, and commercial institutions. They also collected deposits for more than 5000 communication devices from government agencies, delivery services, and private trucking firms.

Early servicing agreements called for a total receiver price of $2900 plus a $35 monthly servicing fee. In the early 1990s Geostar's officials planned to use this and other money raised from private investors to launch three privately owned communication satellites into geosynchronous orbits aboard the French Ariane (Fig. 4.1). Geostar's engineers demonstrated their basic concept using both

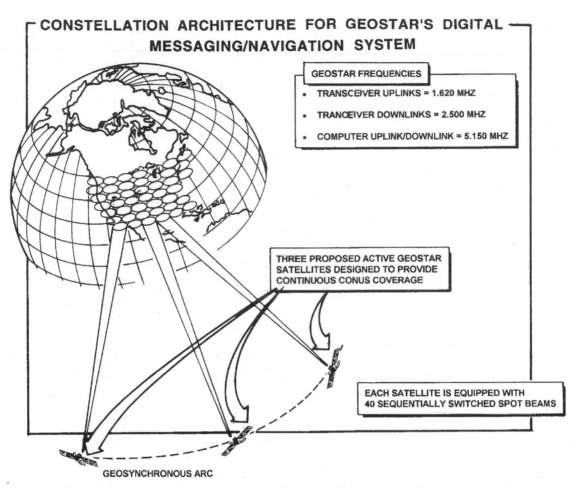

**Figure 4.1**  Geostar's long-range business planners intended to launch three dedicated messaging satellites plus one on-orbit spare capable of taking over if any of the other three failed in space. Uplinks from the Geostar transceivers were rigged to operate at a center frequency of 1.62 GHz with downlink transmissions centered at 2.50 GHz. Uplinks and downlinks to and from the centralized computer facility share the same 5.15-GHz frequency band. A total of 40 sequentially switched spot beams from each of the three satellites work together to provide ample user coverage and frequency reuse.

ground-based transmitters and a "piggyback" payload carried aboard a commercial communication satellite (GStar 2). Later, they obtained approval from the Federal Communications Commission (FCC) for the frequency allocations and the orbital slots they needed for their satellites.

### Gerard O'Neill's Data Relay Concept

The Geostar communication/navigation concept was developed by Dr. Gerard O'Neill in 1978. His inspiration was born of tragedy when two airplanes slammed together above sunny southern California. "In clear skies over San Diego, the nation witnessed a horrible air tragedy," wrote Rob Stoddard in

*Space Communications.* "Despite good weather conditions and state-of-the-art communications, a small private plane collided in midair with a crowded 727 commercial jetliner, sending scores of people plunging to their deaths." O'Neill, himself a frequent air traveler, was shattered by the early descriptions of the incident even before he learned that a close friend was aboard the 727. "There had to be a way to avoid this type of tragedy in the future," he later told a reporter.

By funneling his grief into constructive channels and harnessing his considerable scientific expertise, O'Neill developed and perfected the Geostar concept, a satellite-based radionavigation/communication system that provides precise position coordinates and allows users to exchange simple "telegram messages" anywhere in the United States.

Unlike the Navstar receivers, which perform their own navigation solutions, Geostar's receivers are more akin to simple two-way radios. When a Geostar receiver requests its position, interrogation pulses are automatically transmitted to all three of the geosynchronous satellites, which in turn relay the request to a centrally located computer on the ground (Fig. 4.2). The ground-based computer then performs the navigation solution and sends the results back to the user through one of the three satellites. For a small fee, the same communication channels can be used to relay short "telegram messages" between two or more Geostar subscribers.

Geostar's communication links are more complicated than the ones that service the Navstar users. But the receivers are simpler, and the mobile population O'Neill and his colleagues chose to serve, real-time communications services, are far more valuable than position-fixing services.

## Early Financial Arrangements

By December 1985, Geostar had received firm orders for at least 5000 transceivers, each secured by a 5 percent deposit against the full purchase price of $2900 each. According to *Aviation Week,* Sony and Ma/Comm were each given contracts to deliver 10,000 Geostar receivers. Geostar officials also developed plans for more sophisticated pocket-size receivers to cost about $500 each. Most of their early orders came from transportation companies, including Mayflower Leaseway Transportation and W.R. Grace. Four government agencies also purchased large numbers of Geostar receivers.

Several years ago, Geostar's researchers conducted a series of tests at a rugged site in the Sierra Nevada Mountains to check the operation feasibility of the Geostar concept. Later a piggyback payload carried aboard a communication satellite, GStar 2, demonstrated that the hardware also worked in space.

For proper operation, the full Geostar system required 51 separate inventions. In November 1982, the first patent was developed, and by October 1986 the FCC cleared the way for the Geostar Corporation to launch the necessary satellites into space. Over 50 individuals aired comments at the hearings, which included supporting testimony from the Association for American Rail-

# THE GEOSTAR NAVIGATION SYSTEM

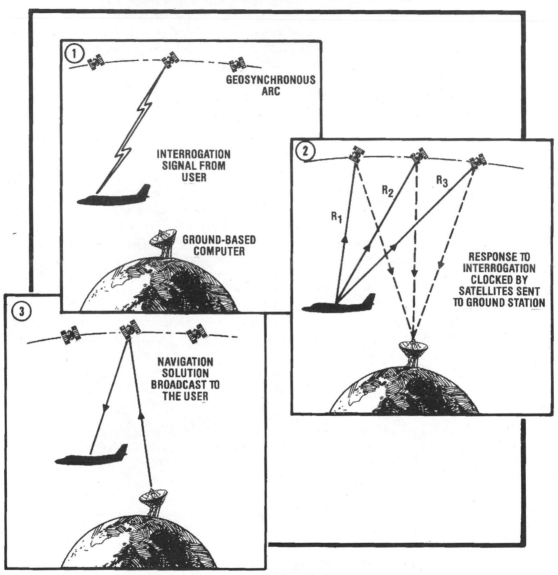

**Figure 4.2**  Unlike the Navstar Global Positioning System, Geostar's architecture calls for a large, ground-based computer to solve for each user's position whenever a request is received. Once the solution is complete, the appropriate position coordinates are to be sent back to the user through the satellites. This approach involves more complicated communication links and requires a costly mainframe computer at "mission control," but it simplifies the design of the receivers and provides the capability for the exchange of two-way "telegram messages." The Geostar Corporation, a private, profit-making company, raised tens of millions of dollars in funding for this space-age project before it went bankrupt.

roads, the Bureau of the Census, several local police and fire departments, the Federal Drug Enforcement Administration, and two airlines, Western and United.

By the end of the third quarter of 1985, Geostar officials had raised $12.5 million from equity placements at $1.25 apiece. "It started off being individual investors," said Geostar's director of finances. "But in the meantime, it has evolved over time to being fairly significant institutional investors." However, despite impressive success in capital markets, Geostar managed to raise only a small fraction of the money it needed to launch the complete constellation of dedicated satellites. The documents submitted to the FCC forecast total on-orbit installation costs at just under $277 million. Overall expenditure and revenue projections further indicated that Geostar expected to raise $1.99 billion over a 7-year period, much of it from receiver sales and services to old and new clients.

## The First Worrisome Hints of Future Difficulties

Unfortunately, an on-orbit failure soon occurred when a 45-lb piggyback payload for Geostar was carried into orbit aboard GTE's GStar 2 satellite. The payload, which was built by the Astro-Electronics Division of RCA, operated successfully for about 6 weeks before it failed. Geostar officials filed an insurance claim for its full value of $2.7 million and began making preparations for their next launch.

The second serious setback occurred when the shuttle *Challenger* exploded. In that same era there were also several expendable booster-rocket failures. Like many other companies with payloads waiting on the ground, Geostar officials found that reliable launch services were not easy to obtain. Their officials held exploratory talks with representatives from China and Japan, but they still expressed confidence that the French Ariane would be able to carry their payloads up to the geosynchronous altitude.

The company was doing reasonably well until FCC officials authorized two other firms to begin servicing Geostar's primary position/messaging markets. Another form of competition also began to come online from mobile communication satellites which will, at a considerably higher price, provide two-way voice communications for trucking fleets and enormous numbers of other land-mobile users.

Gerard O'Neill's fundamental concept was technically sound, but it may have been too advanced for its time. By 1991 Geostar's managers had run into serious difficulties with their company's financiers. Soon they were forced to fold up the business and sink into receivership.

## Mobile Datacomm's Planned Recovery of Geostar's Assets

In 1993, a new company called Mobile Datacomm, headed by former executives from the Comsat Corporation, approached the FCC for permission to offer low-cost satellite-based mobile positioning and data relay services for the lower 48 states.

They planned to acquire the unused assets of Geostar and begin offering a service to locate and track moving vehicles, send messages, and transmit digital data to and from small vehicle-mounted receivers. Eventually they also hope to perform similar services for small, hand-held receivers.

The Comsat Corporation had previously purchased the necessary equipment from Geostar before company officials decided not to get into the remote data services business. Consequently, Geostar's mobile communication equipment modules are available for purchase if the FCC approves the deal.

Mobile Datacomm officials are planning to lease data relay capacity on the Spacenet 3 and the GStar 3 satellites, which they will access from their new hub station in Clarksburg, Maryland. Target markets include federal government bureaus, air-related users, and large and small maritime vessels. Commercial trucking firms will also receive some of their attention.

## OmniTRACS' Mobile Communication Services

Until recently many long-haul truck drivers were forced to wrestle their big rigs into remote truck stops two to four times each day so they could phone in verbal status reports. Small wonder they became irritated when, as often as not, harassed dispatchers were forced to put them on hold.

Each telephone stop typically burned up 45 minutes or more during which the driver's productivity—and often his pay rate, too—dropped to zero! Not surprisingly, driver turnover rates were exceedingly high. At some trucking companies, they sometimes approached, or even exceeded, 100 percent per year. Fortunately, in 1987 Qualcomm, with headquarters in San Diego, California, brought to market the OmniTRACS system, which yields a far more efficient method for regular driver reporting, especially for large trucking concerns operating in the lower 48 states (excluding Alaska and Hawaii).

The OmniTRACS messaging system, which relies on orbiting satellites and highly capable minicomputers on the ground, helps dispatchers manage large trucking fleets with maximum practical efficiency. OmniTRACS marketeers offer their customers vehicle position reporting, digital message exchange between truck and dispatcher, and frequent digital readouts detailing current vehicle performance and load-status information.

So far OmniTRACS terminals have been installed in about 40,000 vehicles including 18-wheel tractor-trailers, minivans, marine vessels, automobiles, and even a few bicycles! In a typical day, 450,000 Ku-band messages are processed at San Diego, each consisting of about 70 alphanumeric characters.

## Remote Tracking and Fleet Dispatching

OmniTRACS is a digital spread-spectrum system that employs CDMA modulations and frequency-hopping techniques. It uses GTE's GStar 1 and GStar 4 geosynchronous communication satellites for message relay in the Ku frequency band (12 to 14 GHz). GStar 1 and GStar 4 are domestic communication satellites positioned over the equator at 82 and 103 degrees west longitude (south of Florida and Texas, respectively).

The OmniTRACS hub station in San Diego picks up digital spread-spectrum messages from the various motor vehicles using a bent-pipe messaging system as illustrated in Fig. 4.3. The position of each vehicle is determined by measuring the signal travel times associated with precisely timed pulses that are sent up from the San Diego hub through the two satellites and back down to the vehicle. The transceiver in the vehicle measures the time delays, then transmits a digital report back to the hub station where the vehicle's longitude and latitude are automatically calculated. The OmniTRACS positioning solutions are accurate to within 1000 ft. The system is also rigged to employ Navstar GPS positioning for improved accuracy and for coverage in regions of the world where two-satellite triangulation coverage is either inaccurate or nonexistent.

The digitally encrypted messages are picked up at the San Diego hub station (network management center) by a 25-ft dish antenna. The positioning pulses are handled separately by a dish antenna a little more than 12 ft in diameter.

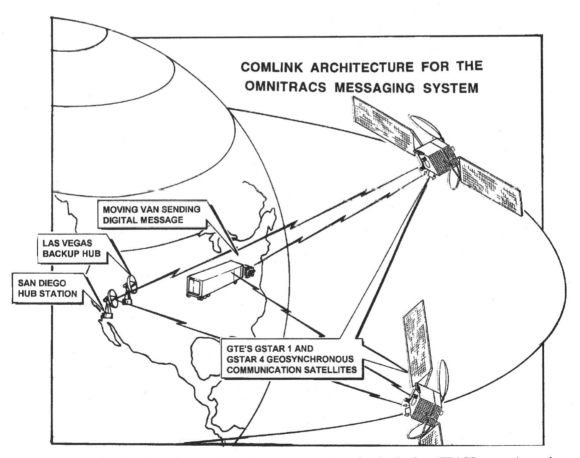

**Figure 4.3**  Large trucking companies constitute the primary target market for the OmniTRACS messaging and position location system, which currently employs the GStar 1 and GStar 4 satellites for continuous message relay. The hub station in San Diego, California accesses the digital pulse trains from the satellites and distributes incoming messages to the intended recipients using conventional telephone links.

## USING THE NAVSTAR SIGNALS TO NAVIGATE YOUR FAMILY CAR

In 1959, the 3.25-lb Vanguard satellite, America's "beeping grapefruit," accidentally demonstrated one form of spaceborne radionavigation when researchers at Johns Hopkins University realized that its orbital motion created Doppler shift variations that differed sharply depending on the location of the ground-based observer. That intriguing observation eventually became the basis for the spaceborne Transit Navigation System, which is still being used today to fix the positions of ships at sea. But today's more modern Navstar navigation receivers determine their geographic positions by measuring the signal travel times of precisely timed pulses coming down from semisynchronous satellites.

Visitors at the 1982 World's Fair in New Orleans got an opportunity to see an unusually interesting exhibit on the Navstar navigation techniques in their future.

"Have you seen that fancy navigation display at the Chrysler Pavilion? It puts little maps on a television screen right in the dashboard, with an electronic marker that shows where you are."

"I saw it. And I loved it! But I hear they're not really selling it. At least not yet."

True, the dashboard-mounted device on display was not yet for sale. In fact, it was not even fully functional. Chrysler's engineers had fibbed a bit when they put their display together. But in the meantime automobile companies all over the world have begun to market units that guide ordinary private cars to their destinations. And in Orlando, Florida you can drive an Avis rental car rigged with electronic Navstar navigation.

Signals from the Navstar satellites fix the car's location, which is displayed against a background of the local streets on the television screen. In the most popular systems, your car's marker remains at the center and, as you drive, the electronic map translates and rotates so that the street ahead always points toward the top of the screen. Look out the windshield, or down at the display and you will see exactly the same scene displayed in exactly the same way.

Compact discs, similar to the ones that carry Whitney Houston's voice into your home with such vibrant realism, store color-coded maps for all 50 states. Touch a button and the map scale will instantly change to any of seven different levels of magnification. The biggest scale displays the entire 41,000-mi interstate highway system. Zoom in for the tightest closeup and all the streets in your immediate vicinity will gradually pop into view.

---

The steerable antennas mounted in the vehicles are vertically polarized for both send and receive. Those on the two geosynchronous satellites are both horizontally and vertically polarized.

Many of the short telegram messages are transmitted automatically from sensors attached to the trucks in the fleet, but individual drivers can also send and/or receive digital telegram-style messages of their own. Most of these messages are formatted automatically as the driver presses the keys on his keypad to fill in the blanks on preformatted electronic forms. A typical trucker sends or receives seven messages in a hard day of driving, not including the automatic position-fixing "messages" that are automatically activated by the Omni-TRACS system once per hour.

Free-form messages can also be sent to and from the big rigs. A large-screen video display at dispatching headquarters shows where all the vehicles in the fleet are currently located, where they came from, and where they are going next. Color coding of the icons allows the vehicles to be grouped by fleets, regions, or zones. If desired, certain messages can be directed toward them all at once as a group.

A "pan and zoom" feature allows the dispatcher to zoom in on a specific vehicle landmark. The dispatcher can also pan across the country for a bigger picture. Weather reports, traffic condition alerts, and instructions to drivers for load pickups all help modern trucking companies operate at or near peak efficiency.

Current truck status information helps dispatchers reduce dead-head miles, detect off-course vehicles, and minimize time spent at truck stops, thus improving gas mileage, lowering maintenance costs, and reducing idling intervals.

An extra optional paging system allows truckers who are waiting for a new consignment to avoid hourly call-ins. The driver can be up to 1000 ft from his vehicle and get some rest or go to sleep knowing for sure that the system will send an alert when his load is ready for pickup and delivery.

### The Hardware Units Carried Aboard Each Company Truck

The OmniTRACS communication system installed aboard each vehicle in the fleet includes three basic types of hardware units:

1. A ruggedized portable control-display unit with keypad and backlit liquid-crystal display

2. A communications terminal housing the electronics modules

3. A continuous-tracking antenna in a sealed dome

Figure 4.4 shows how these three units are mounted in the cab and the engine compartment and on the truck. The antenna, which is usually positioned on top of the cab, is an automatic tracking Ku-band horn in a sealed Lexan radome. An electric motor directs the antenna toward the satellite no matter how the truck twists and turns as it travels along the road. A swiveling antenna of this type is mechanically complicated, but it permits the highest possible data rates, the most error-free transmissions, and the highest possible message reliability. The antenna unit weighs approximately 11 lb. The communication unit and the control-display unit weigh 16 and 2.5 lb, respectively.

### OmniTRACS' Vehicle Information System

Many of the trucks being served by the OmniTRACS messaging system are outfitted with Vehicle Information System (VIS) modules that monitor the current status of the vehicle and automatically generate electronic summary reports. As Fig. 4.4 indicates, the various VIS modules are mounted in the dashboard, in the transmission, inside the engine compartment, and in some cases, on the trailer rigs.

Outputs from the VIS sensors are combined with the position reporting system outputs to help trucking company executives monitor and manage the operation of each individual truck—and all of its companions in the entire fleet. The VIS sensors measure and report such critical operating parameters as engine revolution rates, intake air temperature, oil temperature and pressure,

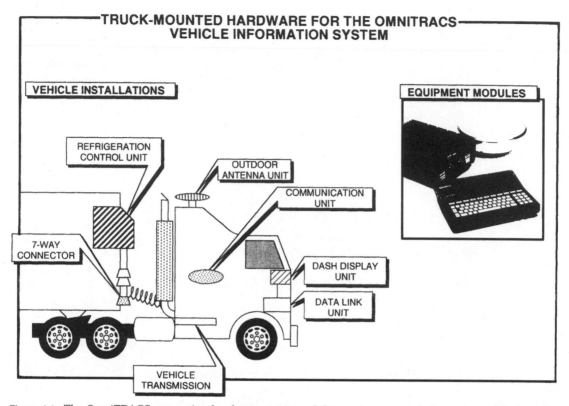

**Figure 4.4** The OmniTRACS messaging hardware consists of three interconnected elements: a sealed aerodynamic dome housing the steerable antenna, a ruggedized communications transceiver, and a user-friendly control-display unit. These three units are cable-connected to one another. They may also, if desired, be connected to an optional Vehicle Information System, which monitors various vehicle parameters such as engine revolutions per minute (rpms), oil pressure and temperature, coolant temperature, and vehicle speed. (*Photograph courtesy of Qualcomm, Inc.*)

and coolant temperature. For some refrigerated trucks the sensors also monitor the current behavior or the trailer's refrigeration equipment.

To illustrate the practical value of diagnostic data on real-time vehicle performance, consider a specific case in which an electronic fuel injector has failed. A dash light warns the driver, and, almost simultaneously, the OmniTRACS messaging system transmits a fault code through one of the GStar satellites to the San Diego hub station, which in turn relays the same information to the trucking company's dispatcher and maintenance experts. These experts quickly assess the severity of the problem and phone ahead to a remote repair facility where the necessary spare parts are gathered up in preparation for the trucker's arrival a half-hour later.

### Truck Routing and Load Matching Techniques

One important function of the OmniTRACS software routines is to automatically match vehicles with the proper power levels to the loads they are assigned to carry for both full-load and partial-load multistop operations. These computer processing routines allow dispatchers to optimize freight selection and to maximize capacity utilization for the entire fleet.

The software routines also facilitate load solicitation. If the system indicates that a half-dozen trucks are to be emptied in Atlantic City on the same day, company solicitors can begin contacting shippers in that area to book more loads.

Another OmniTRACS software routine helps dispatchers identify beneficial load-swapping opportunities. This routine (Fig. 4.5) figures out when two or

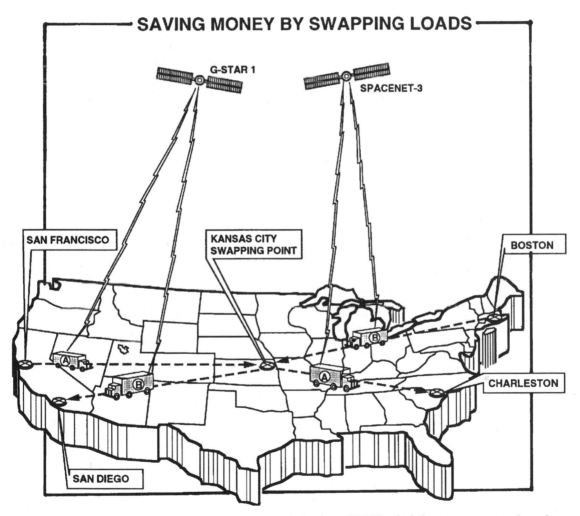

**Figure 4.5**  The software routines used in conjunction with the OmniTRACS vehicle location system seek out beneficial opportunities for load swapping that might, otherwise, be overlooked. In this particular example, two trucks from the east and the west coasts pause in Kansas City, Kansas to swap loads. The two drivers then head back toward their respective coasts. Load swapping provides many economic and humanitarian benefits. The drivers especially like it because it allows them to get back to their families much quicker than they otherwise would.

more company vehicles will be passing through the same location at approximately the same time. It then checks to see if swapping their loads at that location could help improve vehicle utilization, driver time at home, or customer satisfaction.

## Satellite-Based Paging Services

A total of 5.6 million business travelers are potential customers for the wide-area satellite-based paging systems now passing into common use. Current market penetration is only about 3 percent of that total, so the door is wide open for explosive growth—which is already beginning to take place. "Satellite-based paging systems have been growing at a phenomenal rate," notes industry observer George Lawton. "Skytel, the nation's largest satellite-paging network, has experienced annual growth of 128 percent since 1988."

As Fig. 4.6 indicates, the Skytel system relays paging messages through commercial communication satellites and back down to FM radio stations which then piggyback the digital pages onto their normal FM broadcasts. The digital paging messages are modulated onto the FM sidebands so they do not interfere with the station's normal signal.

Electronic pagers come in three different flavors: the most inexpensive models issue simple beeps telling the client to phone a special 800 number to pick up a message. The other two types send numeric messages (the telephone number of the party who called) and alphanumeric symbols (a short "telegram" message with useful instructions).

Leased paging channels on commercial communication satellites are quite inexpensive because so little bandwidth is required. A simple binary page can be transmitted with less than 50 bits. Skytel leases only a single 4800-bit/s satellite channel to handle over 155,000 paging subscribers. But such paging channels do not actually need even that much capacity. "You can cover half a million subscribers at 4800 bits per second if you have only numeric users," observes Jai Bhagat, vice-president of Skytel.

Future expansions may require a bit more communication capacity. Some of today's paging services are rigging their pagers so they can drive electronic mail (E-mail) systems and fax machines. A 100-word E-mail message requires about 3200 bits, and a one-page compressed fax might require 16,000 bits or more.

Paging services have been switching from localized terrestrial links in favor of spaceborne transmission techniques because the paging industry has been moving from a local operation to regional as their best revenue-producing customers venture over larger portions of the country and the world. The spaceborne approach also simplifies the communication links between a network of transmitters scattered over a large geographical region.

Pat Gearty, vice-president of engineering at the American Paging Corporation, is tremendously impressed with the economic benefits of paging through space. "The advantage is that American Paging rents satellite channels for $4000 a month and the costs do not multiply whether we link 10 transmitters

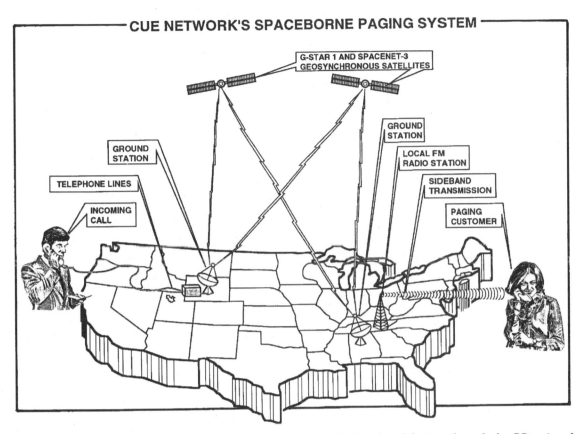

**Figure 4.6**   The Cue Network sends its Skytel paging messages with digital modulations through the GStar 1 and Spacenet 3 geosynchronous communication satellites and back down to widely dispersed FM radio stations. When an FM station picks up a page, it rebroadcasts it as a sideband modulation, which does not interfere with the station's usual scheduled broadcast. Most of today's pagers are small, light, and unobtrusive units available for a few hundred dollars each.

or 1000," he notes. "With other means, such as leased lines, you are talking $50 a month for each line."*

Paging messages relayed through orbiting satellites directly to the paging units may also soon reach the commercial marketplace. Inmarsat, for instance, is instituting a paging service in which digital messages will radiate from their new breed of geosynchronous communication satellites. However, Inmarsat's pagers are rather large—about the size of a brick. And some experts are pessimistic about the ability of Inmarsat's signals to penetrate buildings and other large urban structures.

---

*Pat Gearty was quoted in George Lawton, "Paging: The Mouse That Roared," *Satellite Communications,* September 1992, pp. 29–32.

Motorola's Iridium project, which is slated for initial operation in 1996, will also offer direct-broadcast paging capabilities to their clients. The Iridium handsets will likely be rigged for separate paging services if the client does not wish to pay a premium for voice. The Iridium satellites will be launched into low-altitude orbits, so the paging units will probably be smaller and the satellite signals will have more penetration power to pass through foliage, buildings, and other structures.

## Cargoes that Phone Home

On any given day, 5 million cargo shipping containers are on the move throughout the world. They ride on railroad cars, on flatbed trucks, and on the decks and in the holds of large riverboats and oceangoing cargo ships. While it is in

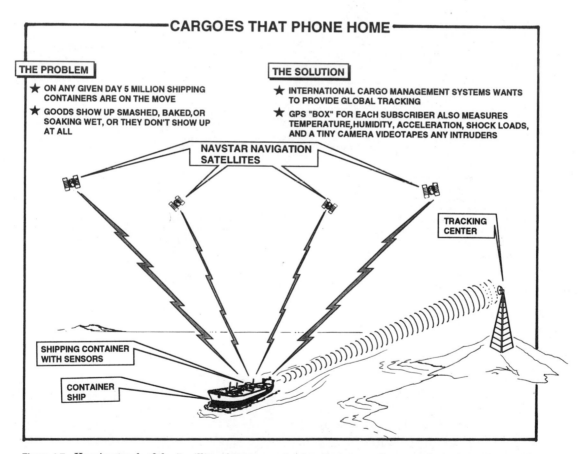

**Figure 4.7**  Keeping track of the 5 million shipping containers that are on the move throughout the world every day is not an easy assignment. But spaceborne digital data links coupled with the position-fixing capabilities of the Navstar GPS satellites can provide an inexpensive solution for tomorrow's shippers and their clients. Digital pulse trains reveal the status and location of each container being tracked. In some cases, a tiny video camera attached to the transceiver even videotapes any intruder who might attempt to open or enter the shipping container to steal the valuable cargo inside.

transit, a shipping container is extremely hard to track, and when it shows up at the other end of its journey the goods inside may be soaking wet, bent, damaged or broken, coated with dust or dirt, or smashed to smithereens. Sometimes the cargo inside does not show up at all.

Fortunately, digital pulse trains relayed through orbiting satellites can be used to track hazardous cargoes and shipping containers as they move along toward the proper loading dock. As the sketches in Fig. 4.7 indicate, a miniature Navstar GPS receiver periodically fixes the location of the shipping container and relays that data to the tracking center's computers through communication satellites orbiting high overhead. At any time a worried client can dial a special 800 number to determine the current location and status of any shipping container.

The transceiver attached to the shipping container also contains a miniature GPS chip set (positioning receiver with no display) and a radio transmitter together with various sensors designed to monitor the status of the cargo nestled inside. The sensors measure and record temperatures, humidities, acceleration, shock loads, and the like. Some fancy versions may even be rigged with a tiny television camera that will videotape any intruder who attempts to open or enter the shipping container.

# 5

# Satellite Messaging for Commercial Jets

*I think it most unlikely that aeronautics will ever be able to exercise a decisive influence on travel. Man is not an albatross.*

H. G. WELLS
*British science fiction writer, 1901*

In 1957, when the Russians launched the first Sputnik, I was a junior in college chasing pretty coeds on a lush green campus in the bluegrass region of Kentucky. That next Friday afternoon, I hitchhiked 60 mi along a dusty country road to my hometown, Springfield, Kentucky, population 2000. Early the next morning, I was walking down Main Street in front of Milburn's Shoe Repair Shop, when I ran into an old boyhood buddy, chubby, freckle-faced Johnny Hardin, who, as usual, had a big grin on his face.

"Hey Tommy," he said, "the Russians just launched a new artificial earth satellite. They have one, do you think we should have one?"

I'm ashamed to admit it now, but at the time I couldn't think of a single reason why our country should hurl an artificial satellite into space. But, before I could say anything, Johnny Hardin answered his own question: "I don't think we should be willing to spend all that money. After all, there's *nothin'* out there!"

Johnny Hardin, it turns out, was precisely right, there is nothing out there. In fact, there are three kinds of nothing out there—no gravity, no air, nothing to block our view. Neither Johnny Hardin nor I could have imagined it at the time, but three kinds of nothing are today making all our lives richer and more abundant in a hundred dozen different ways.

Thirty-seven years have passed since Johnny Hardin and I had that conversation in front of Milburn's Shoe Repair, and during that time hundreds of artificial satellites have been launched into outer space where they take full advantage of the three different kinds of nothing waiting there.

Communication satellites have become the biggest and most profitable part of space industrialization, bringing in about $16 billion each year. Today, we are on the brink of another revolution now that mobile communication satellites are making it possible for all of us to keep in touch wherever we may be: on land, at sea, in the air, and even in outer space.

Today, whenever supermarket owner Johnny Hardin boards a commercial jet bound for an international marketing convention, he can call ahead for reservations—while he is in the air! High above the Atlantic or Pacific, his voice travels from his seat up to an orbiting satellite and back down to the ground again. When he comes back home to Springfield, Kentucky he does not linger in front of Milburn's Shoe Repair. Several years ago Mr. Milburn shuttered the business, but the building he occupied is still standing on Main Street right beside the Pontiac Garage. The new owner sells satellite dishes to the local residents who use them to pick up several dozen channels of cable television coming into Springfield from all around the globe.

---

## THE SHOT HEARD 'ROUND AUNT EFFIE'S CABBAGE PATCH

It was the middle of March, but still the ground was covered with fresh snow and the wind swept in over the north pasture and swirled around the gangling apparatus. He flipped up his rough collar against the wind, but it was hopeless; even fastening all the snaps on his galoshes and buttoning the bottom button of his topcoat would not have kept out the chilly Massachusetts wind. He glanced out at the hazy horizon and then up at the launch apparatus hoping he had thought of everything. The test conditions were far from ideal. The cold air could crack the nozzle and even if it got aloft the wind could drive his awkward little vehicle into the ground before burnout. But it was pointless to consider the risks now, he was committed. The launch would take place today.

He posed for a quick photograph and then, crouched behind a wooden lean-to, he cautiously pointed a blowtorch in the direction of the ungainly framework. In an instant, the tiny rocket hurled itself 41 ft into the air and within 2.5 seconds the terrifying roar was over.

It was 1926, Charles Lindbergh had not yet made his transatlantic flight, and yet Dr. Robert Goddard stood over the remains of his tiny rocket, smoldering and unimpressive in the snow, and dreamed of rocket flights to the moon and beyond.

There would be other launches far more impressive. Forty years later, television newsman Walter Cronkite would desperately brace himself against the windows of his trailer as they rattled from the blast of a rocket 3 mi away; but here today in Aunt Effie's cabbage patch, the world's first liquid-fueled rocket had been flight tested.

---

## Today's Terrestrial Links for Airborne Messaging Systems

America's first airborne telephone systems, which are still in operation, do not relay their signals through orbiting satellites. Instead, their calls are transmitted from the plane down to special antennas on the ground, which route them into the conventional switched telephone network linking the United States and the rest of the world.

Airfone systems are easy to use. When your plane reaches its cruising altitude, you merely insert any major credit card into a slot to release the cordless telephone from its cradle as illustrated in Fig. 5.1. Some models are positioned

## MAKING TELEPHONE CALLS ABOARD
## COMMERCIAL JETLINERS

**Figure 5.1** Western Union's portable Airfones have gained enthusiastic acceptance among busy businessmen abroad commercial airline flights. A credit card releases the portable cordless phone so the user can take it to his or her seat for privacy and comfort. Today's calls are being routed into the conventional telephone networks through relay stations located on the ground. But Western Union officials are now expanding and refining their system by relaying Airfone conversations through orbiting satellites.

next to your seat; others are wall-mounted near the restrooms, in which case you can carry the cordless handset back to your seat for a quick chat with family or friends. Transmissions from the phone are picked up by a long antenna loop running overhead along the length of the cabin. They are then broadcast down to antennas on the ground. When you return the telephone to its cradle, your credit card is automatically released. The terrestrial Airfone system (and its major American competitors) charge about $2 per minute for a typical call.

Today's terrestrial Airfone systems work well when they are within transmission range of their ground-based relay stations. But over vast stretches of land and water, coverage does not exist. Fortunately, Inmarsat's geosynchronous satellites are available, at a premium price, to fill in the coverage gaps.

## Satellite-Based Message-Relay Techniques

"Three international consortiums that offer satellite telephone services to the aviation industry are doing a hard sell to press recession-stung airlines to equip their fleets with satellite equipment," says airline industry observer Robina Riccitiello. The three firms currently supplying voice messaging equipment—Skyphone, Satellite Aircom, and Globalink—argue convincingly that the world's major airlines should be quite willing to invest in the new technology.

According to their line of reasoning, every airplane can become a "flying money machine" capable of pulling in solid revenues from phone calls, faxes, and other airborne communication services.

Skyphone executives, for instance, are projecting annual incomes from satellite phone calls and digital data services that could top $400,000 for each airplane in the fleet. With customer charges reaching as high as $10 per minute on some international flights, cash-strapped airlines are eagerly exploiting this promising new way to keep cash money flowing in. The airborne equipment is expensive now, but, as more and more airborne phone systems are installed, per-plane costs should drop to more affordable levels.

## Domestic and International Aircraft Installations

A few hundred airplanes are already equipped with Inmarsat voice systems, with as many as 2000 expected to come online by decade's end (Fig. 5.2). Each installation currently costs somewhere between $300,000 and $500,000. Equipping 2000 airplanes will thus represent a total investment of nearly $1 billion for the decade of the 1990s.

Why are the world major airlines willing to make such enormous investments when industry profits have been so scarce? The necessary capital expenditures are, indeed, quite large, but payoff intervals are often a matter of only a few months.

## Tapping Today's Airborne Revenue Stream

"Singapore Airlines has been averaging about 20 passenger calls per flight, each averaging about 2.5 minutes, for which the caller pays $8.80 per minute,"

# PROJECTED SATCOM INSTALLATION RATES FOR COMMERCIAL JETLINERS

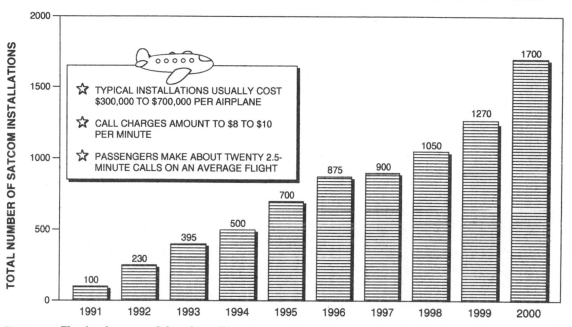

Figure 5.2 The hardware modules that allow commercial jetliners to relay digital data and voice through Inmarsat's commercial communication satellites cost somewhere between $300,000 and $700,000 per plane. But, despite the high cost of the equipment, several hundred international airplanes have already started using the Inmarsat satellites for digital message relay. As these projections indicate, nearly 2000 airplanes will probably be equipped with satcom voice messaging systems by the dawn of the 21st century.

notes Philip J. Klass of *Aviation Week.* Make a few quick calculations, and you will soon begin to understand why Singapore's airplanes are, indeed, turning into "flying cash machines," especially when you come to realize that they have been rigged with only two voice channels each. The newer systems are being rigged with six voice channels bouncing off Inmarsat's geosynchronous satellites. One channel is usually reserved for the airplane's flight crew, but the other five are available to serve passengers with digital data, voice, and fax.

Inmarsat officials are also planning to introduce a new global paging service that will probably be able to reach passengers in flight. Inmarsat's marketeers and airline executives are hoping that paging messages will trigger even more airborne phone calls from international business travelers. One indication of the economic viability of international satellite voice services comes from United Airlines. Recently company leaders decided to outfit their entire fleet of 747-400s and 767-300s with multichannel voice messaging systems. Similar systems will go into their new 777s now on order from Boeing Aircraft.

American air travelers have been using flight phones for more than a decade, served by terrestrial broadcast systems. Soon their European counterparts will have access to equivalent services in their communications-hungry part of the world.

## Europe's New Digital Messaging Techniques

The European system, now in the installation phase, takes advantage of the latest digital multichannel voice-messaging techniques. The European transmissions occupy the L-band frequencies between 1.6 and 1.8 GHz. The first few ground stations, which have a transmission range of about 125 mi, are being installed near Glasgow, in Yorkshire and North Wales, at Lyon, France and Malmo, Sweden, and near Milan, Italy.

The major air carriers of Europe are being charged nearly $5 per minute for their airborne phone services. Of course, carriers will undoubtedly add a hefty surcharge of their own. So passengers who use the system will have to pay even more. The Europeans' much higher rate compares with the $2 per-minute charge paid by airline passengers in the United States.

## Inmarsat's Hardware Installations

The earliest airborne telephone systems installed in the United States had a single voice channel linking the aircraft to the ground. Consequently, impatient passengers sometimes had to wait their turn to use the phone. To break the bottleneck, some of the airlines installed two parallel Airfone systems. The newest spaceborne systems, by contrast, yield six usable channels: five for the passengers, one for the crew. Multichannel communication systems of this type currently cost $600,000 to $700,000 each.

The sketches in Fig. 5.3 highlight the key characteristics of a typical antenna system installed aboard a 747 jumbo jet. Notice how the two conformal phased-array antennas are installed on the two opposite sides of the fuselage angled upward toward space. The external portion of each of the two antennas is less than 0.3 in. thick. Such a thin profile offers only a small amount of drag to the slipstream flowing over the aircraft.

The Ball AIRLINK high-gain antennas sketched in Fig. 5.3 are normally integrated with the Collins SAT-906 suite of avionics equipment. This integrated digital system provides multichannel voice and data transmission capabilities. In addition, a small blade-type low-gain omnidirectional antenna is usually installed atop the fuselage for low-rate data communications from the cockpit to air traffic control. Inmarsat officials are predicting that the organization's revenues from airline telecommunications will grow from $1 million in 1992 to $45 to $50 million by 1996. Over the past few years, Inmarsat has invested about $60 million in developing the necessary technology for airline communication systems.

Four Inmarsat 2 satellites are currently in orbit over the equator and four Inmarsat 3s are being constructed under a $330 million contract. The new Inmarsat 3 models are equipped with multiple spot beams and an agile frequency reallocation scheme to provide greatly improved and more versatile communication services. Inmarsat's current customer base of 31,000 users is expected to mushroom to 70,000 by 1997. The Boeing Commercial Group estimates delivery of 12,000 new passenger and freighter aircraft between 1992 and 2010, so the airborne communications systems will probably continue to grow.

# A COMMERCIAL JETLINER INSTALLATION FOR A TYPICAL SPACEBORNE VOICE-MESSAGE RELAY

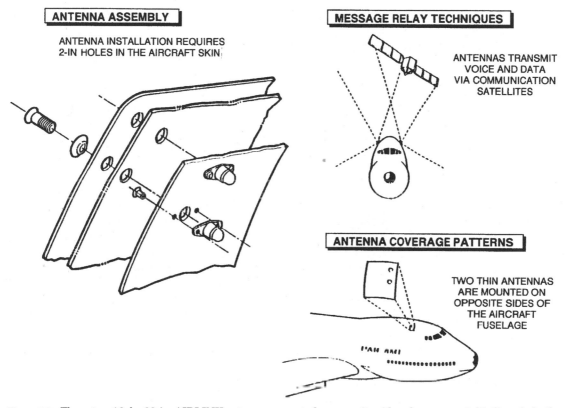

**ANTENNA ASSEMBLY**

ANTENNA INSTALLATION REQUIRES 2-IN HOLES IN THE AIRCRAFT SKIN

**MESSAGE RELAY TECHNIQUES**

ANTENNAS TRANSMIT VOICE AND DATA VIA COMMUNICATION SATELLITES

**ANTENNA COVERAGE PATTERNS**

TWO THIN ANTENNAS ARE MOUNTED ON OPPOSITE SIDES OF THE AIRCRAFT FUSELAGE

**Figure 5.3**  These two 16- by 32-in. AIRLINK antennas mounted on opposite sides of a commercial jetliner help the passengers and crew maintain voice communications, even when the plane is flying over the Atlantic or the Pacific Ocean. The two conformal phased-array antennas are steered electronically to send voice messages to and from Inmarsat's geosynchronous satellites. The antennas are flush-mounted with the aircraft's skin, so they do not create appreciable drag or interfere with the slipstream's flow.

## Passenger Benefits

"You're over the Pacific, on your way to Hawaii. Oops, you forgot to ask the neighbors to pick up your mail! No problem, just pick up the phone in your United Boeing 777, and give them a jingle." Jane Reller published that rousing description in a 1993 edition of *Challenge Magazine.* Her enthusiasm for the operational ease of modern satcom systems is entirely justified. And, best of all, the system she described is available today!

Many airline passengers claim that one of the biggest benefits of being on an airplane is that they can escape the insistent distraction of telephones. But, once they arrive at their cruising altitude, with a handy handset at their fingertips, they suddenly find a hundred dozen different reasons to make that call. Studies have shown that passengers check stock prices, reserve rental cars, and even send bouquets of roses while they are in the sky. They also plan marketing strategies with business representatives, leave long, rambling messages on voice mail systems, remind their colleagues to pick them up at the airport, and sometimes they even whisper "sweet nothings" to wives and sweethearts and lonesome dogs and cats, too!

## Benefits to the Airlines

Airplanes flying over large, well-traveled land masses, such as Europe or the North American continent, are tracked by radar and vectored toward their destinations by ground-based radionavigation systems. But until recently airplanes flying high above the oceans have been inaccessible to ground-based aiding techniques. Their pilots used voice-reporting, instead, with periodic reports transmitted over high-frequency radio links.

All communications are in English, but accents cause problems and so do misunderstandings and garbled communications. Position reports come in only every hour or so and, unfortunately, the high-frequency communication channels tend to be overcrowded, unreliable, noisy, and inaccessible. High-frequency voice channels are also sometimes knocked out by charged particles created by intense sunspot activity.

Most of today's flight plans are laid out before the plane takes off, and because of communication difficulties during long ocean crossings, they are usually not revised enroute except in fairly serious emergencies. Even if a request for a change is made, it is not always honored by the professionals at air traffic control. "Often a controller will reject a requested change in a flight plan," says *Aviation Week* reporter Philip J. Klass, "because of uncertainties as to the position of potentially conflicting traffic—which typically make their position reports only once every hour."

"At least 5 percent of our operating costs on average is now wasted because of growing traffic density and current traffic-control constraints," says William B. Cotton, manager of air traffic systems at United Airlines. "Because crews are very nervous about not being able to get their flight plan altitudes, they will add 10,000 lb of extra fuel to the flight—6000 lb of which will be burned just to carry the extra fuel!"

Now that Inmarsat's messaging services are widely available, the international airline companies have been experimenting with real-time digital data requests to update their flight plans while they are in the sky. Thus, for example, if a European Airbus encounters stiff headwinds, the flight crew can request a less choppy and more fuel-efficient cruising altitude.

According to Clarence A. Robinson, Jr. of *Signal* magazine, Federal Aviation Administration (FAA) experts are convinced that their new digital Data Link System, with its spaceborne communication channels, can "change the way air traffic control is managed from predeparture through landing, dramatically improving flight safety and efficiency and potentially saving the aviation industry billions of dollars annually."

The digital Data Link System is so important to efficient long-range flights because nearly all of today's radio-frequency voice circuits are filled to capacity during air traffic "rush hours" in the sky. According to the FAA, traffic controllers are sometimes talking to 25 airplanes over the same frequency. This is equivalent to a situation in which a large number of people are trying to use the same telephone line all at once but, of course, it is much more dangerous at a cruising speed of 520 mi/hr and an altitude of 37,000 ft.

Safety is not the only important issue driving the airlines toward the Data Link System. One airline executive estimates that his company spends more than $300 million annually due to excessive communication-related delays. With the Data Link System, transoceanic pilots will be able to report their positions automatically every 5 minutes using simple and efficient digital pulse trains. A number of other types of messages such as winds-aloft measurements will also be reported over the same digital transmission links. Lost or garbled messages will be all but eliminated because every incoming message will be automatically acknowledged by the computer on both ends.

Much of the electromagnetic message clutter will also go away. In a series of experiments at Chicago's O'Hare International Airport and at the Dallas–Fort Worth Airport, technicians who were testing the Data Link System for predeparture message exchange found that message traffic declined by 60 percent compared with airports that use voice for the same function. Flight-control experts at the ICAO (International Civil Aviation Organization), the international counterpart for the FAA, are forecasting savings of $5 billion per year if the satellite-based system they are proposing for flight vectoring and aircraft control is implemented on a global basis in keeping with the schedule they recommend. Their approach, incidentally, involves the use of several different types of orbiting satellites including communication satellites, international weather satellites, and satellites devoted to spaceborne surveillance coupled with the position-fixing capabilities of the Soviet Glonass and the Navstar GPS.

Recently a team of researchers at Japan Air Lines conducted a test in which they beamed cloud cover imagery from a meteorological satellite directly into the cockpit of a Boeing 747. This approach provided "the latest cloud cover far better than the usual weather maps," concluded Peter Ishikawa, flight operation vice-president for Japan Air Lines. Company executives at Japan Air Lines are now trying to figure out how to provide similar cloud cover imagery to all the airplanes in their fleet that operate on transoceanic routes.

## Coupling Cockpit Communication with Navstar Navigation

In the summer of 1983, engineers and technicians at Rockwell International's Government Avionics Division demonstrated the airborne navigation capabilities of the Navstar GPS when they sent a Sabreliner business jet across the Atlantic from Cedar Rapids, Iowa to the Paris Air Show. Two Navstar receivers were carried onboard for real-time guidance and control. The pilot made intermediate landings at Burlington, Vermont; Gander, Newfoundland; Reykjavik,

## HUNTING FOR DINOSAURS WITH THE NAVSTAR SATELLITES

The Navstar navigation satellites can be used to guide airplanes toward worldwide destinations, but their L-band signals can be used for more imaginative purposes, like trying to find the best way to make friends with extinct creatures. Consider the adventures of Dr. Roy P. Mackal, a research biologist from the University of Chicago, and his colleague Dr. Herman Regusters, an engineering consultant at Cal Tech's Jet Propulsion Laboratory. During a visit to the Congo in the early 1980s, Mackal interviewed more than 30 pygmies who pumped him up with vivid descriptions of a strange, hippopotamus-size beast they called the *mokele mbembe*.

According to their account, the giant animal, which spent most of its time immersed in swampy waters, had a "reddish-brown body, a long neck, a relatively small head, and a long, powerful tail." When Mackal showed them drawings of various animals, they all agreed that their *mokele mbembe* was most like the brontosaurus—the largest animal ever to walk the earth.

Mackal became even more interested in coming back for a careful search when he learned that several Western observers had also picked up traces of the oversized animals deep in the jungles of the Congo. In the 1770s, for instance, French missionaries had reported seeing "the tracks of a clawed animal the size of an elephant," and about two centuries later, "two separate German expeditions hinted at the existence of the elusive monster."

When he returned to the Congo in 1982, Dr. Roy P. Mackal carried video cameras and sonar devices designed to monitor the underwater movements of the *mokele mbembes*. His aim was to find them paddling around in the swollen streams during the rainy season when water levels would likely rise enough for their families to venture out of their hiding places in the jungle and come downriver in search of food.

Mackal had planned to carry a Navstar receiver to help him locate the swampiest regions as revealed by false-color images obtained from the Landsat earth resources satellites. He tried to borrow a receiver from the Department of Defense, but unfortunately all available units were tied up in making military tests. He went to the Congo anyway, but found no *mokele mbembes*. The conclusion thus seems obvious: if only a Navstar receiver had been available, public zoos throughout the world would probably now have in their display cages extended families of brontosauruslike creatures supposedly long since extinct.

Iceland; and London before touching down at Le Bourget Field outside Paris. After he landed the plane, the pilot used the GPS positioning signals to taxi to a presurveyed spot on the apron of the runway, ending up within 25 ft of his intended destination.

Rockwell's Sabreliner business jet thus became the first airplane to cross the Atlantic using spaceborne navigation all the way. To many airline professionals, that flight also highlighted the practical potential for using satellite signals for flight vectoring and air traffic instructions. Advantages usually cited include improved safety, large fuel savings, better passenger comfort, and more efficient use of transoceanic and continental airspace.

More than 250,000 airplanes are currently operating in the United States. These include approximately 3500 commercial airliners carrying 270 million paying passengers to produce $30 billion in annual revenues. They land and take off from about 7000 domestic landing strips, only 12 percent of which are equipped with electronic landing aids. Nearly 10,000 air traffic controllers help maintain safe separation distances and, in other ways, attempt to ensure the safety and efficiency of America's gangling and often inefficient air transportation system.

## Dependent Surveillance Techniques

In November 1992, Norway's air traffic controllers began using GPS positioning receivers to guide helicopters to offshore oil platforms on the North Sea. Precise positioning data from the satellites were relayed from the helicopters to the Norwegian air traffic controllers through Inmarsat communication satellites. This simple demonstration foreshadows revolutionary developments in air traffic control.

The emerging approach hinges on *dependent surveillance techniques* for keeping track of large, but well-equipped, fleets of commercial planes. Instead of observing each plane visually or tracking it on a radar screen, the airplane itself is required to reveal its own position. Signals from the Navstar satellites (and their Soviet counterpart called Glonass) fix the airplane's position, which then broadcasts its position coordinates in digital form through other orbiting satellites.

With a properly implemented automatic dependent surveillance system, flight crews always know where other nearby planes are located. This allows them, to some degree, to select the most fuel-efficient routes and altitudes, rather than being nailed down to preselected airways. The dependent surveillance approach will also permit the FAA and its international counterparts to reduce the required separation distance between aircraft, thus increasing fuel efficiency and the passenger-carrying capacity of the overall system. Members of the Radio Technical Commission for Aeronautics (RTCA), a nonprofit advisory board, are convinced that these improvements will generate $13.2 billion in airline benefits between 1995 and 2015.

## Supplementing the GPS Constellation with Geosynchronous Satellites

To further enhance the capabilities of the dependent surveillance technique, my colleagues and I at Rockwell International devised a concept for supplementing the coverage of the GPS constellation with additional satellites at the geosynchronous altitude. These extra satellites were designed to provide both navigation and communication to airborne users. Each extra satellite would be rigged to broadcast precisely timed navigation pulses together with real-time communication messages streaming between flight crews and air traffic control personnel.

Three geosynchronous satellites spaced 120 degrees apart along the equator would be able to provide global coverage except for the extreme northern and southern latitudes. A fourth dual-function satellite provides redundancy to back up any of the other geosynchronous satellites that might fail. Orbital maneuvering thrusters allow one or more of the functioning satellites to be repositioned for improved coverage. All of the new geosynchronous satellites are rigged with communication links to help disseminate air traffic control commands and current information on the health status and the integrity of the other satellites in the constellation—including the ones at the geosynchronous attitude and the ones in 12-hour semisynchronous orbits.

The satellite transmitters and the transmitters onboard the airplanes employ frequency-division multiple access and time-division multiple access messaging techniques with precise time slots synchronized by the timing signals from the GPS satellites. The digital FDMA and TDMA modulation techniques greatly increase the capacity of the allotted frequency spectrum assigned to air traffic control.

### Three-dimensional Air Traffic Control

If the dependent surveillance technique is to be successfully implemented, every airplane in the sky (excluding hang gliders and ultralights) must be equipped with a GPS receiver capable of broadcasting the airplane's current position, course, and speed. The appropriate multidimensional state vector is transmitted to one or more of the geosynchronous satellites for rebroadcast to the appropriate air traffic controller and to other, nearby planes.

The video display format proposed by the Rockwell team is presented in Fig. 5.4. This particular electronic display is specifically designed for the air traffic controllers. A similar but less elaborate pictorial representation is displayed in the cockpit of each plane so each pilot in the sky has access to essentially the same information.

The air traffic controller views the local terrain on his or her video screen in the form of a tilted checkerboard pattern with crisscrossing lines of longitude and latitude. An inverted translucent cone is associated with each airplane. The apex of the cone touches the local terrain at the longitude-latitude location directly below the aircraft. The height of the cone represents the altitude of the aircraft, and the diameter of its base (which is at the top) represents 3 minutes of travel at the airplane's current speed.

If an air traffic controller wants to communicate with the pilot of a particular plane, he or she merely touches its icon on the screen with a finger, and the computer automatically sets up the proper communication channel. When contact has been established, the appropriate cone automatically changes color. Nearly all routine communications are accomplished by means of the touch-screen buttons running around the edges of the controller's video screen. To send a message, such as "descend to 15,000 ft," the controller touches two or three buttons, checks the message on the screen, then touches the send button.

Narrow-band digital pulse trains carry the message to the receiver onboard the airplane where they are routed through electronic voice synthesizer chips to produce high-quality stereo sound in the pilot's headphones. The message is simultaneously displayed on the pilot's video screen. If the pilot wants to hear the message again he or she merely touches one touch-screen button for a stereo repetition.

### Landing and Taxiing Operations

During landing maneuvers, the airplane's video screen switches to the special display mode sketched in the upper left-hand corner of Fig. 5.5. During landing

# AIR TRAFFIC CONTROL DISPLAY

**Figure 5.4** Tomorrow's air traffic controllers will probably employ three-dimensional satellite-driven computer displays so they can keep track of the planes occupying their sectors of the airspace. In this conceptual system, an inverted cone is associated with each plane. The apex of the cone, which is at the bottom, touches the local terrain at the proper longitude-latitude location. The height of the cone is proportional to the altitude of the plane, and the diameter of its base represents 3 minutes of airborne travel. Touch-screen technology coupled with voice-synthesizer chips help foster clear and efficient digital voice communications between the pilots and their assigned air traffic control personnel.

the pilot sees a "picket fence" on the video display and, as the airplane travels down the glideslope, the rails on the picket fence rapidly melt away.

Once the airplane is on the runway, the display screen switches to an overhead "bird's-eye view" of the airport with a circle surrounding each plane (Fig. 5.5). Larger circles indicate faster taxiing speeds. The plane on the right does not have a corresponding circle because it is temporarily parked on the runway.

A taxiing system similar to the one depicted in Fig. 5.5 is already being tested at Chicago's O'Hare International Airport. Proponents of this approach are eventually planning to have the computer-based system track every vehicle that roams around the runway, including luggage vans, security vehicles, and food trucks, too.

# DISPLAYS FOR TERMINAL MANEUVERS

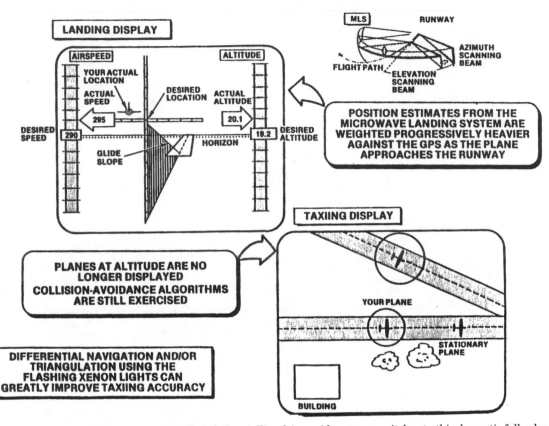

**Figure 5.5** During landing maneuvers, the pilot's satellite-driven video screen switches to this dramatic full-color "picket fence" method of display. As the airplane approaches the runway, the individual rails in the picket fence systematically slip from view. The taxiing display (lower right) shows a bird's-eye of the airport with a circle surrounding each plane automatically scaled to its speed. The airplane on the right is sitting still, so no circle is shown.

## Takeoff and Landing Using Carrier-Aiding Techniques

America's forward-thinking aviation officials have been conducting a series of tests to determine whether the GPS (and the Soviet Glonass) satellites are sufficiently accurate and reliable to guide commercial jetliners and other airplanes during takeoffs and landing operations.

These operations were supposed to be handled at large, well-equipped airports by the new Microwave Landing System, but FAA officials have recently expressed willingness to consider abandoning that system in favor of competing spaceborne navigation if the efficacy of the new spaceborne navigation systems can be demonstrated.

Ground equipment for the Microwave Landing System is slated to cost $2.6 billion, with another $1.7 billion to be ponied up by the major airlines to install the necessary avionics modules aboard their planes. A decision concerning which approach to pursue is scheduled for 1995, after a series of tests comparing the capabilities of the two techniques.

One test series is being conducted by a group of research engineers at Stanford University (Fig. 5.6). Their engineering approach employs relatively inexpensive "pseudosatellites" designed to enhance the accuracy of the Navstar navigation solutions. A pseudosatellite (false satellite) sits on the ground and transmits navigation signals similar to the ones coming down from the Navstar satellites.

Most Navstar receivers execute their positioning solutions by measuring the signal travel times of the Navstar pulses as they move from the satellites to the ground.

However, a more accurate technique uses the L-band carrier waves being transmitted by the satellites to enhance the accuracy of that solution. Carrier-aided solution techniques of this type were pioneered by professional surveyors. Unfortunately, a surveying-type solution usually requires about 45 minutes of measurement time. This long-term measurement interval is necessary

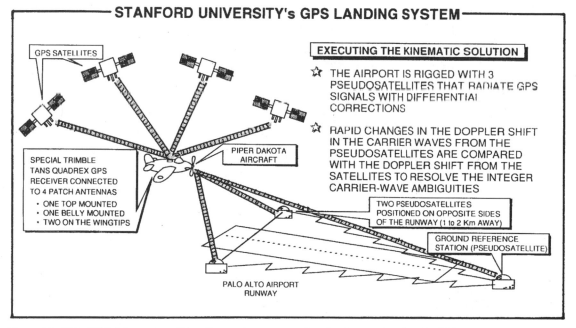

**Figure 5.6** This GPS-driven aircraft landing system, now under development at Stanford University, uses carrier-aided solution techniques to enhance the navigation of the conventional Navstar satellite signals. Rapid changes in the Doppler shift from the carrier waves broadcast by the pseudosatellites on the ground help the aircraft computer resolve the intrinsic ambiguities in the carrier-aided solution. The wingtip-mounted antennas take advantage of another carrier-aided solution technique to help determine the airplane's attitude as it comes down toward the runway for landing.

to allow the Navstar satellites to travel along an appreciable arc across the sky. As a satellite moves along its orbit, Doppler shift changes in its electromagnetic carrier waves give the computers enough information to help alleviate solution ambiguities. The ground-based pseudosatellites help the computers perform a similar service much quicker for landing operations. As the airplane moves toward the runway, its motion creates the necessary Doppler shift with respect to the pseudosatellites located on the ground.

The attitude angles of the airplane can also be determined by employing carrier-aided solution techniques. This is possible using two separate antennas mounted on its wingtips. The difference in the path length of the parallel signals coming down from the Navstar satellite to its wingtips provides information on the instantaneous tilt angle of the airplane's wings.

# Part 3

# Mobile Satellite Constellations Now Emerging from the Drawing Boards

# 6

# Selecting the Proper Constellation Architecture

*Television won't last. It's a flash in the pan.*
MARY SOMERVILLE
*Radio technologist and pioneer, 1948*

Over the past few years, communication satellites in space and cellular telephone systems on the ground have both enjoyed spectacular success. Even during the recent recession, when the U.S. economy was largely stagnant, spaceborne communication systems consistently posted growth rates of 20 or 30 percent, and, during that same era, cellular telephone revenues shot up at an even faster pace. Today 3 million Americans have purchased satellite dishes and 15 million of their fellow citizens own and operate cellular telephones. Moreover, as prices continue to spiral downward, millions of other ordinary private citizens are poised to join tomorrow's telecommunications revolution.

These two success stories have motivated creative engineers to devise new ways to combine communication satellites with cellular telephone technology. Such a partnership has, in fact, already begun to form now that geosynchronous satellites are beginning to serve relatively heavy and bulky tetherless telephones installed aboard oceangoing vessels, long-haul trucks, and commercial jetliners.

In parallel with these terrestrial and spaceborne developments, the miniaturization of electronic components has been moving forward at a remarkable pace, so that today millions of tiny electronic processing circuits can be crammed onto a computer chip hardly larger than a baby's fingernail. Microminiaturization techniques are resulting in smaller satellites and smaller transceivers on the ground.

The pieces have definitely been falling into place, so is it now possible to construct pocket telephones that will connect people everywhere with crisp, clear messages relayed through orbiting satellites? Many engineers are convinced that they can package earth stations into tiny *Star Trek*-sized personal communicators. Some venture capitalists are willing to invest huge amounts of time and billions of dollars to make it work. Of course, they are also hoping to reap billions of dollars of profits in return.

The constellation architectures today's aerospace engineers have been proposing are amazingly diverse, and so are the markets they are struggling to serve. Some of them have been aiming toward digital data services using small constellations of satellites. Others are focusing their attention on store-and-forward communication concepts in which the satellite records a digital message, then pumps it down to a distant recipient at some other point along its orbit. Some proposed constellations are intended to handle voice messaging services with bigger antennas, more transmitter power, and broader bandwidths to accommodate the increased information. These systems will, in essence, provide global cellular telephone services beamed down from outer space.

## Constellation Selection Trades

The next four chapters will discuss emerging concepts for tomorrow's mobile constellations soon to be providing tetherless communication for planet earth. Industry experts have, so far, proposed and evaluated three distinctly different constellation architectures:

1. Geosynchronous constellations hovering in the sky at the 19,300-nautical mi altitude
2. Low-altitude constellations with swarms of satellites barely skimming over the earth
3. Constellations of satellites in medium altitudes a few thousand miles high

The choice of orbital altitude is driven by three specific constellation characteristics which are discussed in detail in the next three sections:

1. *The Orbital Environment.*   This section includes discussions of such altitude-dependent characteristics as the Van Allen Radiation Belts, spaceborne solar illumination, signal time delays for the electromagnetic messages, and the anticipated elevation angles of the satellites as viewed by users on or near the ground.

2. *The Estimated Cost of the Proposed Constellation.*   The overall cost of a mobile communication constellation is strongly influenced by various altitude-dependent characteristics including the number of satellites required to provide global coverage, the estimated cost of constructing and launching each satellite, and the projected lifetime of the average satellite once it reaches its orbital destination.

3. *Cost and Complexity Assessments for the Overall System.*   The total life-cycle cost of a mobile communication system centers around a number of altitude-dependent characteristics. These include the necessity for hand-offs to new satellites during the average conversation and/or the necessity for crosslinking the messages from satellite to satellite. Other important issues include cost and com-

plexity estimates for the ground control segment, and the cost and weight of the personal communicators. The possibility of achieving incremental usage while the constellation is being constructed can also be an important cost driver. If incremental usage is possible during construction, the resulting revenues can be used to help pay for the remaining satellites in the constellation.

Many industry experts have developed forceful opinions concerning the best orbital locations for tomorrow's constellation of mobile communication satellites. Unfortunately, those strongly held opinions tend to clash with one another. That is why at least a dozen different companies are planning to launch their satellites toward such a wide variety of different destinations.

The altitudes for today's proposed constellations, in fact, differ by a factor of 50. Their orbital inclinations differ by more than 100 degrees. And the size of the proposed constellations varies from one single satellite hovering along the geosynchronous arc to more than 800 of them whizzing over the surface of the earth, barely clearing the atmosphere.

Many of the proposed constellations are intended to provide digital data relay rather than voice. However, the discussions to follow will concentrate on the much more sophisticated voice messaging satellite systems. One other important point should be strongly emphasized. Today's large geosynchronous communication satellites achieve frequency reuse through spatial diversity. Each narrow-beam antenna on the ground transmits a specific frequency toward a particular geosynchronous satellite with an electromagnetic beam narrow enough to avoid illuminating its neighbors in space. Consequently, that same frequency can be reused by other ground-based antennas accessing other geosynchronous satellites.

Spaceborne mobile communication systems achieve frequency reuse by adopting a completely different approach. They do not point their antennas toward specific satellites. Instead, most employ *omnidirectional* antennas with coverage patterns spanning the entire sky. The various mobile users distinguish the messages directed toward them by CDMA, TDMA, or, in some cases, FDMA. These modulation techniques allow large numbers of users to achieve frequency reuse on a grand scale.

## The Orbital Environment

In choosing their constellation architectures, aerospace engineers must bear in mind that the space flight environment their satellites will experience varies enormously with orbital altitude. Van Allen radiation damage, solar illumination intervals, and signal travel times, in particular, are strongly dependent on spacecraft altitude. So are satellite elevation angles as seen from the earth, and the constellation's exposure to hazardous man-made space debris.

### The Van Allen Radiation Belts

An orbiting satellite is exposed to the energetic charged particles in the Van Allen Radiation Belts and to the ones streaming down from the sun during solar storms. These energetic particles can cause logic upsets in a satellite's electronic circuit chips, damage some of its delicate components, and reduce the power-generating capacity of its solar arrays.

Spacecraft shielding and careful positioning of the most vulnerable components can help mitigate these detrimental effects. So can the use of radiation-resistant electronic devices such as gallium-arsenide solar cells. Unfortunately, these approaches complicate the overall spacecraft design and increase costs by a substantial amount.

As Fig. 6.1 indicates, the Van Allen Radiation Belts consist of two donut-shaped regions containing energetic protons and electrons traveling around the earth's magnetic field lines in complicated spiraling trajectories. The lower belt, which is composed of a rich mixture of protons and electrons, reaches its peak intensity at an altitude of about 2000 nautical mi. The upper Van Allen Radiation Belt, which is composed primarily of electrons, reaches its peak intensity at 10,000 nautical mi.

## THE VAN ALLEN RADIATION BELTS: DOSE RATE VERSUS ALTITUDE

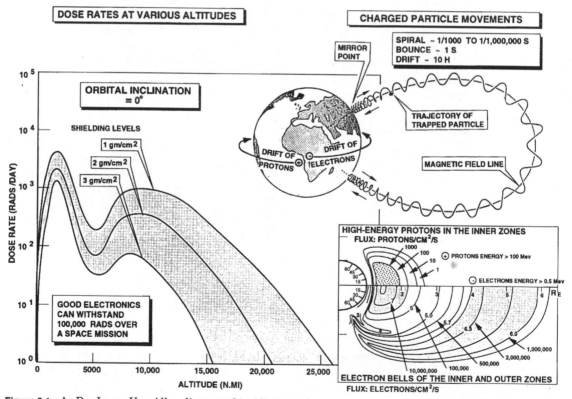

**Figure 6.1**   As Dr. James Van Allen discovered in 1959, two donut-shaped radiation belts encircle the earth. The lower belt is a rich mixture of energetic protons and electrons. The upper belt is composed primarily of electrons. Peak intensities occur at 2,000 and 10,000 nautical mi, respectively. The energetic particles in the Van Allen Radiation Belts can damage delicate spacecraft components, cause logic upsets, and decrease the efficiency with which the satellite's solar arrays can generate electrical power.

Low-altitude satellites fly under the most intense portions of the lower Van Allen Belt, but they experience additional radiation due to the South Atlantic Anomaly, a region between Uruguay and the southern tip of Africa where the earth's magnetic field dips down close to the earth. The South Atlantic Anomaly is caused by the mismatch between the magnetic axis and the geographic axis of the spinning earth. Low-altitude satellites crossing through the South Atlantic Anomaly experience much higher radiation levels than they would at the same altitude above any other portion of the globe. Atomic oxygen in the earth's upper atmosphere also damages some of the surface components of low-altitude satellites.

Geosynchronous satellites orbit above the most damaging portions of the upper Van Allen Radiation Belt, but they experience more damage from intense solar flares than either low-altitude or medium-altitude satellites. This is true because geosynchronous satellites are well above the best shielding effects of the earth's magnetic field.

The graph in Fig. 6.1 shows how the radiation dose-rate (rads/day) varies with increasing orbital altitude. Notice that the low-altitude constellations and the geosynchronous constellations usually experience less Van Allen radiation than those in medium-altitude orbits. Notice also that the best medium altitude is about 5000 nautical mi above the surface of the earth. A satellite at that level lies between the upper and the lower Van Allen Radiation Belts.

## Solar Eclipse Intervals

The eclipse periods and the eclipse durations are strongly dependent on the orbital altitudes and the orbital inclinations of the individual satellites in a spaceborne constellation.

During portions of its on-orbit mission, a low-altitude satellite may spend 35 percent of its time or even more in the earth's shadow, with shadowing intervals occurring 13 to 15 times each day. When it enters the earth's shadow, a satellite may stay in darkness for as much as 35 minutes before it reaches full sunlight again. During those repeated day-night intervals, the satellite's thermal control devices and its electrical storage batteries are subjected to repeated stress cycles. Such stress cycling tends to create design penalties and shorten the satellite's on-orbit life.

As Fig. 6.2 indicates, both the maximum duration spent in the earth's shadow and the frequency of these outages are considerably more manageable for medium-altitude and geosynchronous constellations. A geosynchronous satellite spends about 99 percent of its on-orbit life in full sunlight and those in medium-altitude orbits spend at least 90 percent of their time illuminated by the sun.

## Signal Time Delays

Electromagnetic waves travel at a constant speed, so sending a voice message through a geosynchronous satellite takes longer than sending it through a low-altitude or a medium-altitude satellite. "When satellites were first introduced for voice service, telephone companies were very concerned about customer

## SPACECRAFT ECLIPSE INTERVALS FOR VARIOUS ORBITAL ALTITUDES

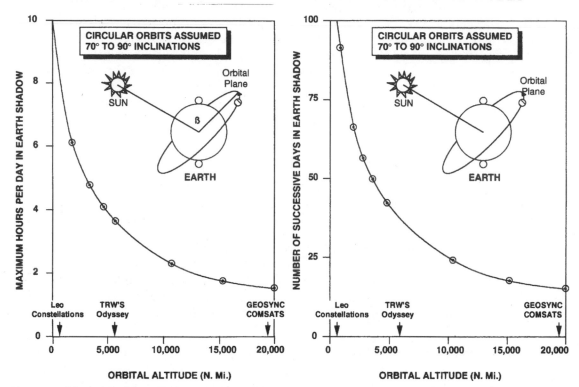

**Figure 6.2**  A low-altitude satellite enters the earth's shadow during almost every orbit and, throughout long portions of the year, it spends 30 to 35 percent of its time in the earth's shadow. Geosynchronous and medium-altitude satellites, by contrast, are illuminated by the full perpendicular rays of the sun 90 to 99 percent of the time. Shadowing durations and shadowing frequencies are extremely important to spacecraft engineers because a satellite cannot generate electrical power while it is being shadowed by the earth. Frequent shadowing intervals also create severe thermal control problems for spacecraft design teams.

reaction to echo, delay, and sometimes crosstalk," says TRW's mobile communications expert Roger J. Rusch.

These marketing concerns led Bell Telephone Laboratory researchers to conduct a series of experiments in which they purposely introduced time delays of various durations into simulated phone conversations to study customer reactions. Soon they had reached a few simple and useful conclusions: "Time delays produced talk overlap or confusion if the one-way time delay exceeded 0.3 second... for an interaction time between two communicators of 0.6 second," notes Rusch. "But time delay was not perceptible if the time delay was less than 0.2 second."

An electromagnetic signal headed toward a geosynchronous satellite takes about 0.125 second to get up there and another 0.125 second to come back down again. So, in accordance with the Bell Telephone Company results it might seem that the total time delay of 0.25 second would be acceptable to most consumers. Unfortunately, other factors, including terrestrial network delays, voice codec processing delays, satellite crosslinking, and multihop communication relays can all add extra increments to the delay caused by simple signal propagations.

Serious consumer resistance to signal time delays has already been encountered in ordinary international satellite relay systems. Time delays have, in fact, virtually eliminated geosynchronous relay for voice communications *within* the borders of the United States and western Europe. However, roughly 70 percent of international continent-to-continent message traffic still streams through geosynchronous satellites, but, even in that case, underseas fiber optic cables have been nibbling away at market share.

The one-way time delay for terrestrial telephone networks is only 0.005 to 0.020 second due primarily to the digital transmission architecture used in relaying the messages. For transoceanic fiber optic cables the corresponding time delay amounts to 0.10 second. In this case, periodic buffers and intermediate booster relays create most of the delay.

The raw transmission time delay encountered by mobile communication satellite constellations is strongly dependent on orbital altitude. The propagation delay for low-altitude constellations ranges from 0.005 to 0.010 second, depending on altitude and the satellite's position in the sky. For medium-altitude constellations the propagation delay amounts to about 0.070 to 0.080 second. For geosynchronous satellites it reaches 0.25 to 0.27 second.

Unfortunately, in a practical telecommunication system a number of time delays must be added to these simple propagation delays. For example, the computer processing associated with digital compression techniques typically adds another 0.06 to 0.08 second. Processing delays can also be introduced by satellites that relay their calls to ground-based gateways for circuit switching (a common mobile-communication satellite approach). In addition, TDMA systems add additional delays by temporarily storing blocks of information in buffers, while messages are in transit toward the intended recipient.

If a mobile satellite constellation employs satellite-to-satellite crosslinking, the signals may be delayed as much as 0.09 second in each electronic buffer. Some spaceborne transoceanic telephone satellite calls require four separate relays. In this case, a total time delay time amounting to 0.36 second could be introduced by the buffers alone.

Figure 6.3 summarizes the various time delays associated with typical low-altitude, medium-altitude, and geosynchronous mobile communication systems. For comparison purposes, terrestrial communication links and transoceanic fiber optic cables are also included in Fig. 6.3. Low- and medium-altitude constellations encounter the shortest time delays. The time delays associated with the geosynchronous constellations tend to be considerably longer.

## Spacecraft Elevation Angles

Spaceborne mobile communication systems are being designed to supplement terrestrial cellular, but spaceborne systems will not have enough power to penetrate buildings and foliage to the same extent that today's cellular transmit-

## SIGNAL TIME DELAYS FOR VARIOUS MOBILE COMMUNICATION CONSTELLATION ALTITUDES

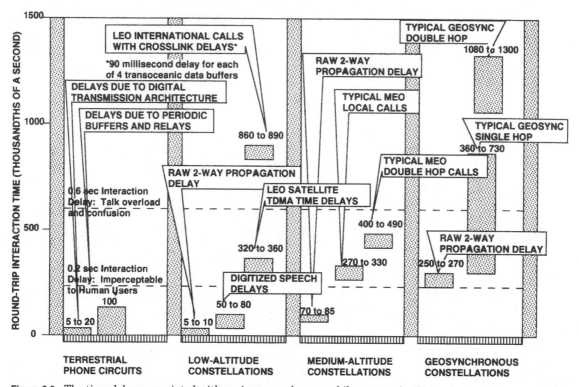

**Figure 6.3** The time delays associated with various spaceborne mobile communication systems are summarized in this figure together with comparative data for short-range and long-range terrestrial communication systems. Low-altitude mobile communication constellations experience the shortest propagation delays followed by the medium-altitude constellations. The propagation delays associated with geosynchronous satellites are long enough to cause relatively serious overlap in conversations between inexperienced users. Long pauses in international phone calls cause particularly severe problems whenever the emphasis of the conversation is on romance.

ters can. When a building or any other obstacle comes between the user and the satellite, the link can be broken. The elevation angles of satellites as seen by observers on the ground are of crucial importance to overall systems design. The line-of-sight elevation angles are strongly dependent on the orbital altitudes chosen for the satellites.

Mobile communication satellites positioned at the geosynchronous altitude are fixed geodetically, so, once a pocket telephone user finds a suitable transmission site without blockages, the satellite will generally remain available for the duration of the call. However, geosynchronous satellites tend to lie close to the horizon for users who live at northern latitudes. A resident of Stockholm, for instance, looking toward a geosynchronous satellite due south of his hometown will see the satellite at an elevation angle only 22 degrees above their horizon. Those who make their calls from mountainous terrain or on the streets of San Francisco, Boston, or New York will likely experience frequent blockages due to trees and structures poking up toward the sky.

The satellites in a low-altitude constellation sweep from horizon to horizon in 20 minutes or less, and during that interval they spend an appreciable fraction of their time fairly close to the horizon. Consequently, blockages from natural formations, buildings, and other structures can create serious problems.

Medium-altitude satellites travel across the sky quite slowly and they have higher elevation angles than geosynchronous or low-altitude constellations. TRW's medium-altitude Odyssey constellation with only 12 satellites provides an average line-of-sight elevation angle of 45 to 50 degrees. With the full constellation properly spaced out across the sky at Odyssey's 5400-nautical mi altitude, at least two satellites are always in view and the minimum elevation angle turns out to be only about 30 degrees.

Odyssey's constellation architecture is designed so that at least two satellites will typically be above the horizon at all times. The computers in the system automatically select the best available satellite and make all necessary hand-offs seamlessly, in much the same way that today's cellular systems handle their hand-offs.

In a series of tests for their planned Project 21, Inmarsat's marketeers became convinced that sophisticated users will probably be willing to tolerate moderate transmission delays. Most of them will, in addition, willingly walk or drive to favorable locations where line-of-sight coverage is available. Inmarsat engineers and various others have also been looking at the possibility of coupling their voice messaging systems with high-penetration paging systems and medium-penetration "call announcement" features to make their design approach more attractive to consumers.

## Space Junk

Over 6 million lb of space junk is currently whirling around the earth, mostly in low-altitude orbits. American military personnel continuously track over 7000 space fragments as big as a soccer ball or bigger. Collisions between a functioning satellite and one of those artificial objects—or any of the thousands of smaller ones as big as a garden pea or bigger—could have catastrophic consequences.

Some computer simulations operated by experts at NASA's Johnson Spaceflight Center seem to indicate that, if the orbiting fragment population triples, successive mutual collisions could create so many fragments that missions into space could become too dangerous for human beings to attempt.

Low-altitude mobile communication constellations will experience by far the greatest hazard from orbiting space debris because most of the debris fragments are barely skimming over the earth at altitudes of a few hundred nautical miles. Moreover, the fragments that cross low-altitude orbits have the highest orbital speeds. Any collision between a garden-pea-sized fragment and a mobile communication satellite would most likely generate explosive energy. Even with an intersection angle of 15 degrees, the energy exchange would be about the same, pound-for-pound, as exploding an equivalent amount of TNT.

---

### THE GROWING HAZARD OF ARTIFICIAL SPACE DEBRIS

Venture out into your backyard with a good pair of binoculars and you will get an opportunity to study the destructive potential of hypervelocity impacts. Many of the biggest lunar craters are rimmed with ragged white spokes formed eons ago when lunar subsoil was hurled radially outward by huge meteorites slamming into the surface of the moon. The average impact velocity of those savagely energetic projectiles has been estimated at 44,000 mi/hr.

Now swivel your binoculars gently across the evening sky, and you can uncover solid evidence for the population explosion of space debris whirling around the earth. Look toward the southwest as the sun begins to sink below the horizon. Four or five times every hour you will spot at least one orbiting object shimmering in the light of the setting sun. Now wait an hour or two and notice the number of stars visible to the naked eye. If the sky is dark and clear, and there are no city lights to create background glare, you should be able to see about 3000 stars—roughly half the number of objects now orbiting in one hemisphere. At this moment, space-age technicians at North American Air Defense (NORAD) are tracking more than 7000 objects whirling around the earth including hundreds of operational satellites, plus a much larger number of spent rockets, dead or dying space vehicles, shrouds, clamps, fasteners, and even Ed White's silver glove, which came off during his first walk in space.

Each year about 800 new objects are added, and roughly half that number plunge back down toward earth. Fortunately, most of them burn up before they impact the ground. Objects as small as soccer balls can be tracked by NORAD radars, but much smaller and lighter fragments can present a hazard to travelers in space. Because of their savage velocities, even a space debris fragment as small as a garden pea can damage an artificial satellite. Experts estimate that the total number of objects of destructive size is at least 20,000, perhaps substantially more.

Sixty percent of the trackable objects in NORAD's inventory have been produced by violent explosions in space, nearly 100 of which are known to have occurred. During the Cold War Russian scientists found a way to blow up enemy satellites. Their "killer satellites," big, deadly sawed-off shotguns, created one quarter of the explosions when they tested their space-age killer satellites against targets they launched into space. Spaceborne explosions also occurred when propellant tanks on American upper-stage rockets suddenly ruptured. A few of these rockets were believed to be dead in space for 3 years or more before they exploded. According to industry rumors, early Soviet astronauts may have further aggravated the space debris problem by tossing garbage out of their manned space stations. NORAD's radar imaging devices are not sensitive enough to make a positive identification, but Russian watermelon rinds may be intermingled with the other spaceborne objects our military technicians have observed streaking across their radar screens.

---

Studies show that the intersection angles can have almost any value, even head-on at 180 degrees!

A moderate hazard also exists for geosynchronous satellites because the geosynchronous altitude is such a popular orbital destination and because the geosynchronous satellites are all crowded into essentially the same orbital plane with a 0 degree inclination angle. However, any collisions that do occur will generally be less energetic than those in low-altitude locations because the geosynchronous objects are moving only about half as fast and because their intersection angles would not likely be much higher than 5 degrees.

The satellites orbiting in medium-altitude constellations will not encounter much space debris because the medium altitudes have never been a very popular orbital destination. However, a few satellites in elliptical orbits do cross through that region of space to create a moderate hazard. These include the

Soviet Molniya communication satellites, several dozen of which have been launched into 12-hour, 63.4-degree orbits with earth-skimming perigees and apogee altitudes in the vicinity of geosync. Intersection angles for the intermediate-altitude satellites could have almost any value, so any collisions that happen to occur would probably be catastrophic for both satellites.

## Estimated Cost of the Proposed Constellation

The cost of installing a communications constellation is primarily a function of the number of satellites required, the cost of building each satellite, and the cost of delivering each satellite to its orbital destination.

For profitable operation, a mobile satellite constellation must of course be maintained for several years, so one other important cost item comes into play: the projected lifetime of the average satellite. Average satellite lifetime strongly influences the frequency with which new ones must be built and transported into space. These three key cost drivers are discussed one by one in the next three subsections.

### The Number of Satellites Required

Satellites in higher-altitude orbits can access a larger fraction of the earth, so fewer of them are required to achieve continuous coverage. At the geosynchronous altitude a satellite has message relay access to 42 percent of the globe—a large skullcap-shaped region with the top of the cap positioned on the equator directly below the satellite. A mobile communication satellite in a medium-altitude orbit can see a bit less of the earth. Each 5400-nautical mi Odyssey satellite, for instance, can access about 30 percent at any given moment. Low-altitude satellites, such as the ones that populate the Iridium constellation at the 413-nautical mi altitude, can access only about 5 percent of the earth any given moment.

The two curves in Fig. 6.4 represent the minimum number of satellites required for single- and double-satellite coverage for circular orbits at various altitudes. The number of satellites being planned by a few of today's mobile-satellite corporations are also shown on the chart.

Low-altitude constellations, ranging in altitude from 300 to 700 nautical mi, require several dozen satellites if they are to provide continuous coverage to all or most of the globe. That is why the Iridium constellation at 413 nautical mi is slated to include 66 satellites and the Starsys constellation at 702 nautical mi includes 24 satellites.

At the geosynchronous altitude five satellites could theoretically yield continuous global coverage—provided they were properly spaced around the globe. Seven satellites could provide continuous *dual* coverage. However, these small geosynchronous constellations would have to be launched with 50 to 60 degree inclination angles so they would not be stationary with respect to users on the ground.

The engineers who design geosynchronous constellations usually place their satellites in equatorial orbits with 0 degree inclination angles. In so doing they purposely choose to sacrifice coverage in the (largely inhabited) regions near

# HOW MANY SATELLITES ARE REQUIRED TO COVER THE EARTH?

"How many satellites are required to provide coverage for the earth?" I asked the bright young students at the first session of one of my orbital mechanics short courses.

"What do you mean by coverage?" one of them shot back. I had purposely left my formulation a little bit vague in hopes of encouraging them to be a bit more creative.

"You are the ones who get to define the word 'coverage'," I told them. "That can be part of your answer. You can continue to think about the solution while I provide you with a little background material."

In the early days of the space program, many aerospace engineers concluded that a minimum of six satellites would be needed to provide continuous global coverage. Most envisioned three satellites in circular, equatorial orbits with three more in polar orbits. Six satellites, it turns out, can definitely do the job. But in the early 1960s John Walker in England constructed a five-satellite constellation that could provide continuous global coverage. Each of the five satellites was to be launched into a 55 degree circular orbit at an altitude of at least 6200 nautical mi.

For a long time Walker's constellation with only five satellites was the smallest known continuous-coverage constellation, until John Draim in Alexandria, Virginia, developed a constellation with only four satellites that could do the same job. His constellation used satellites that were to be launched into inclined elliptical orbits. It took John Draim 7 years to solve the problem using both computers and pencil-and-paper calculations. His four satellites were to be placed in *elliptical* orbits with apogee altitudes just beyond the geosynchronous arc inclined at an angle of 31 degrees.

"How many satellites are required to provide coverage for the earth?" I asked my students once again. They were still pondering their answers while I started to unfold my own.

"*Three*," I told them, "if we are willing to ignore the poles!" They were getting a little more interested now. "*Two* if we are willing to put them high above the earth at infinite altitudes." Little ripples of laughter began to echo through the room. "*One* if we don't require continuous coverage!" Surely that was the smallest possible number that could do the job.

"How about *part* of a satellite?" one of the students suddenly interjected. "How about leasing a transponder on another company's satellite?" What a terrific idea! Why hadn't I thought of that?

"How about *zero* satellites," replied another bright young conceptualizer. "How about reflecting powerful radar waves off the moon to cover the earth?" Surely that is the final answer. Or can someone, somewhere figure out how to do it with an even smaller number?

the North and South Poles. Four geosynchronous satellites spaced 90 degrees apart over the equator can cover a continuous band ranging between 70 degrees north and south latitude. This assumes that some users at extreme latitudes will be willing to use satellites situated near the horizon.

As Fig. 6.4 indicates, constellations positioned in medium-altitude orbits can achieve dual earth coverage with only about 10 satellites or even fewer. But below that altitude, much larger numbers of satellites are required to provide continuous global coverage.

## The Cost of Building and Launching Each Satellite

Satellites designed to operate in the three proposed orbital-altitude regimes are constructed with varying complexities, sizes, and on-orbit weights. The cost of transporting them to their respective orbital destinations also varies over a rather broad range. Consequently, designers encounter some intriguing cost

## THE APPROXIMATE NUMBER OF SATELLITES REQUIRED FOR CONTINUOUS GLOBAL CONVERAGE

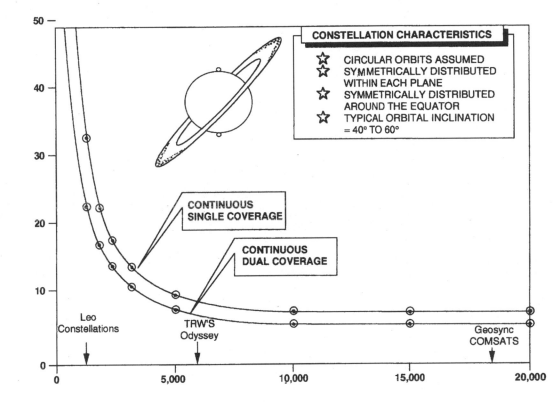

**Figure 6.4** A 500-nautical mi mobile communication constellation requires at least 30 or 40 satellites to guarantee continuous single-satellite global coverage. A medium-altitude constellation (above 6200 nautical mi) could theoretically provide continuous coverage with only five satellites assuming that they are launched into inclined, circular orbits properly distributed around the globe. Dual coverage, in which at least two satellites are constantly visible, typically requires around 40 to 50 percent more satellites at any given spaceborne altitude.

tradeoffs when they study mobile communication constellations headed toward different orbital altitudes.

Geosynchronous mobile communication satellites turn out to be the most complicated and expensive. For effective operation at the geosynchronous altitude, a mobile communication satellite must be equipped with several dozen very narrow spot-beams illuminating adjacent sectors of the ground. Expensive spot-beam design techniques are necessary to allow adequate frequency reuse. This approach is conceptually similar to a cellular telephone system except that each spot-beam in a cellular system is rigged with a separate ground-based transmitter. By contrast, the spot-beams in a communication satellite are produced electronically by the antenna's complicated feed mechanisms.

Because it is so far away from its receivers, a practical geosynchronous satellite must be equipped with very large onboard antennas, complex transponders, and large power-generating devices to serve the small personal communicators situated on the ground below. Of course, the space transportation costs for geosynchronous satellites are much higher, too. Twice as much energy must be expended to launch a geosynchronous satellite to its orbital destination compared with a low-altitude satellite of the same weight.

Low-altitude satellites can be much simpler and cheaper than those positioned at geosync. The free-space propagation losses encountered by an electromagnetic carrier wave are proportional to the square of the distance, so low-altitude satellites can be equipped with smaller antennas and less powerful transmitters. Transporting a low-altitude satellite into earth orbit is also, of course, quite a bit cheaper than transporting one of equivalent weight to a higher-altitude orbit.

If low-altitude constellations did not require so many satellites for adequate coverage, they would be the cheapest possible approach. However, according to Roger J. Rusch at TRW Systems, the intermediate altitudes seem to represent the best compromise between the number of satellites required and the cost of building and launching each satellite into space.

Figure 6.5 summarizes Rusch's results. The three curves on the left correspond to the approximate cost of building a mobile communication satellite bound for different orbital altitudes, the cost of launching it into space, and the total combined cost. Notice that both launch costs and satellite costs rise sharply with orbital altitude. The curve on the right combines the costs for building and launching each satellite with the number of satellites required at various orbital altitudes. The net result is a graph of total constellation cost versus circular orbital altitude.

Notice that the cost reaches a minimum for an altitude in the neighborhood of 3000 to 6000 nautical mi. Smaller constellations can provide continuous global coverage at higher altitudes, but the costs of building each high-altitude satellite and launching it into space are considerably higher. The satellites in low-altitude constellations are lighter, simpler, and cheaper, and easier to launch but unfortunately extremely large numbers of them are required for continuous global coverage. The intermediate-altitude regime apparently represents the best compromise between the cost of building and launching each satellite and the total number that are required.

## Projected Satellite Lifetimes

The on-orbit lifetime of the average satellite in a mobile communication constellation has an important impact on the overall cost of building and maintaining that constellation.

Low-altitude satellites experience frequent shadowing intervals and spend a substantial fraction of their time in the earth's shadow. This puts extra strain on their solar arrays, their batteries, and their thermal control subsystems. Low-altitude satellites also sweep periodically through the South Atlantic

## ROUGH ORDER-OF-MAGNITUDE COSTS FOR BUILDING AND INSTALLING MOBILE COMMUNICATION CONSTELLATIONS AT VARIOUS ALTITUDES

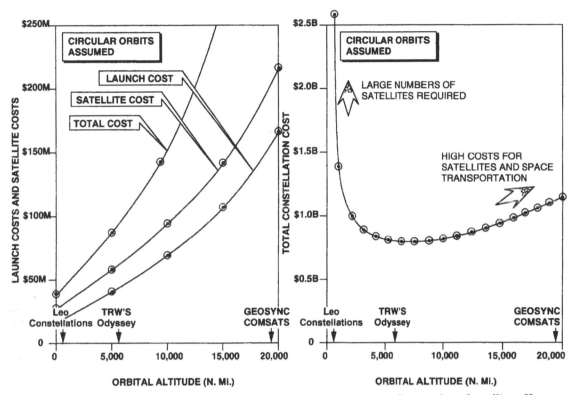

**Figure 6.5** A high-altitude constellation can provide good coverage with a smaller number of satellites. However, building each high-altitude satellite costs extra money, and so does launching it into space. Consequently, it is not obvious which altitude regime yields the least expensive constellation. The curves in this figure represent rough order of magnitude estimates indicating that the lowest-cost constellation has an orbital altitude somewhere in the neighborhood of 4000 nautical mi. The actual optimum cannot be pinned down with a high degree of precision because so many design assumptions are involved. However, these curves probably represent realistic trend lines that can be used in shaping more accurate and detailed investigations.

Anomaly where they are subjected to unusually high levels of radiation from the lower Van Allen Radiation Belt. This reduces the power-generating capacity of their solar cells and damages their onboard electronic components.

Geosynchronous satellites occupy a relatively benign radiation environment and they are not subjected to the radiation in the South Atlantic Anomaly, but they tend to be big and complicated and they must be rigged with high-power electronics and sophisticated multibeam antennas, so their components tend to have fairly high failure rates.

Medium-altitude satellites occupy a more intense portion of the Van Allen Radiation Belts, so they must be designed with extra shielding and radiation-resistant parts. Some proponents of medium-altitude mobile communication constellations are convinced that their satellites will have the longest lifetimes among the ones sent to the three proposed orbital destinations. However, this position is controversial among industry experts.

## Cost and Complexity Assessments for the Overall System

Constellation costs aside, several altitude-dependent considerations drive the overall cost and complexity of a typical mobile communication system. These special considerations include the design features of the ground control segment and the hand-held or mobile personal communicators used in accessing the satellites. These overall architectural considerations are discussed in detail in the next four subsections.

### Are Hand-offs and/or Crosslinks Required?

Each satellite in a low-altitude constellation has instantaneous access to only a small fraction of the earth's surface at any given moment and it sweeps from horizon in 20 minutes or less. For this reason, many of the designers of low-altitude voice messaging systems are planning to provide automatic hand-offs from one satellite to another and/or satellite-to-satellite crosslinks for instantaneous message relay. Crosslinking promotes flexible routing across vast distances. In some cases it also allows for the possibility of bypassing the conventional public-switched telecommunication networks. Unfortunately, hand-offs and crosslinks tend to add complexity to the overall system architecture and increase total costs.

Hand-offs and crosslinks are largely unnecessary for geosynchronous and medium-altitude constellations because the satellites in those altitude regimes have instantaneous access to such large regions of the earth and because the satellites move so slowly across the sky. As seen from the ground, a medium-altitude satellite moves across the sky only about 1 degree/min. So, in a minute its apparent movement is equivalent to the width of one forefinger seen from arm's length.

In order to minimize service disruptions, some low-altitude constellation designers are rigging their systems so users can communicate through two satellites simultaneously. In this so-called "soft" hand-over technique, the personal communicators must be designed to pick up two satellite signals simultaneously and automatically bring them into precise time synchronization. Such an approach helps the user achieve seamless communication service but raises the cost and the overall complexity of the personal communicators.

### Cost and Complexity Estimates for the Ground Control Segment

The simplest ground control infrastructure for a low-altitude mobile communication system requires hundreds of ground stations to access the satellites along the direct line of sight. Crosslinking can reduce the number of stations on the ground, but at the cost of increased satellite complexity. The personal communicators employ omnidirectional antennas, so bore-sight pointing is never required. But most ground control antennas, which are designed to carry vast amounts of information, are required to swivel and tilt to track the satellites as they pass overhead.

The geosynchronous and the medium-altitude satellites cover much larger areas on the earth, so smaller numbers of ground stations are required for the medium-altitude geosynchronous constellations. Dynamic satellite tracking is unnecessary for the geosynchronous constellations because the satellites remain essentially stationary in the sky. Slow-speed tracking is needed for the medium-altitude satellites.

## The Cost and Weight of the Personal Communicators

Low-altitude constellations yield the smallest and lightest hand-held communicators operating at the lowest levels of power. This is true because the satellites in a low-altitude constellation are much closer to their users and therefore, all other things being equal, they can deliver the most powerful signals.

Medium-altitude constellations yield communicators of moderate cost and moderate weight. Future communicators served by geosynchronous satellites are usually projected to be heavier and of moderate cost.

## Can Incremental Services Be Provided?

If incremental services can be provided for certain heavily populated regions of the earth, while the constellation is being installed, revenues from those users can help pay for the construction of the remaining portions of the system. Consequently, any realistic possibilities for incremental services should be carefully evaluated.

The proposed geosynchronous constellations have the highest potential for economically beneficial incremental services. A single geosynchronous satellite can cover 42 percent of the globe—and it is the *same* 42 percent all the time. Thus, a single mobile communication satellite hovering at geosync could immediately start to pull in revenues from Europe, the Orient, the United States, or any other heavily populated target market.

Low-altitude constellations do not have a very promising potential for incremental services judging from the first few that have been launched into earth orbit. Industry studies have shown that cellular telephone customers typically demand usable communication links at least 95 percent of the time. If their coverage intervals drop below 80 percent, they tend to return their cellular telephones and switch to a new communications provider.

The satellites in a medium-altitude constellation will probably have moderate opportunities for incremental services. TRW's engineers, for instance, are convinced that with only 6 of their 12 Odyssey satellites in place, they could provide valuable mid-latitude services in several heavily populated regions of the earth.

Some medium-altitude *elliptical-orbit* satellites have an even greater potential for providing valuable incremental services. With four Molniya satellites launched into 12-hour elliptical orbits, Soviet aerospace engineers routinely provide continuous coverage for the northern hemisphere. With only four more satellites in *inverted* Molniya orbits (with their apogees below the equator) the Soviets could provide coverage for the southern hemisphere too.

## Capsule Summary of the Constellation Selection Trades

Table 6.1 summarizes the various altitude-dependent constellation trades for the three different orbital altitude regimes—low altitude, medium altitude, and geosynchronous altitude. Diagonal shading lines mark the best selections in each category; dot-pattern shading marks the worst.

Of course, finding the best orbital location for a mobile communication system is not as simple as counting the number of shaded rectangles associated with each proposed selection. Each organization is usually interested in servicing a specific target market, so much more detailed studies are required to determine which characteristics are important to that particular clientele and how best to provide the service characteristics they require. However, Table 6.1 is a good starting point in setting up such a detailed study linking available services with customer needs.

## Constellation Architectures Now on the Drawing Board

At least a dozen major companies and consortiums are poised to enter various mobile communication satellite markets. Their efforts will be explored in detail in the next four chapters. Four different types of mobile communication systems will be characterized and evaluated:

1. Mobile communication satellites at geosync
2. Telegram satellites launched into low-altitude orbits
3. Voice messaging satellites launched into low-altitude orbits
4. Medium-altitude constellations

The orbital locations (altitudes and inclinations) for a sampling of the various proposed constellations will be summarized in polar coordinate form in Chapter 11, which also summarizes the salient characteristics of the 13 mobile communication constellations that have so far received the most publicity and attention in popular magazines and technical journals.

**TABLE 6.1  Constellation Selection Trades: Summary Chart**

Legend: ☐ Best Characteristics  ▨ Worst Characteristics

| SELECTION CRITERIA | LOW-ALTITUDE CONSTELLATIONS (~ 400 N. MI.) | MEDIUM-ALTITUDE CONSTELLATIONS (~ 5,000 N. MI.) | HIGH-ALTITUDE CONSTELLATIONS (~ 19,300 N. MI.) |
|---|---|---|---|
| **THE ORBITAL ENVIRONMENT** | | | |
| VAN ALLEN RADIATION | Low levels of radiation in low-altitude orbits. Magnetic field shielding | Moderate levels of radiation in properly chosen orbits | Low levels of radiation in geosynchronous orbits |
| ECLIPSE INTERVALS | Frequent day-night cycling. Satellite in darkness ~ 30% of time | Infrequent day-night cycling. Satellite in darkness ~ 2% of time | Infrequent day-night cycling. Satellite in darkness ~ 1 to 2% of time |
| SIGNAL TIME DELAYS | Shortest signal delay times ~ 0.0C s for 2-way transmissions | Moderate signal delay times ~ 0.1 s for 2-way transmissions | Longest signal delay times ~ 0.25 s for 2-way transmissions |
| SPACECRAFT ELEVATION ANGLES | Rapidly varying elevation angles. Satellites frequently near horizon | Slowly varying elevation angles. Satellites well above horizon most of the time | No elevation angle variations. Satellites near the horizon for high latitude users |
| MANMADE SPACE DEBRIS FRAGMENTS | Large numbers of space debris fragments | Smallest numbers of space debris fragments | Moderate numbers of space debris fragments |
| **ESTIMATED COST OF THE CONSTELLATION** | | | |
| NUMBER OF SATELLITES REQUIRED | Largest number of satellites required. Typically 30 to 60 | Moderate numbers of satellites required. Typically 10 or 20 | Smallest numbers of satellites required. Typically 3 to 6 |
| COST OF EACH SATELLITE AND ITS TRANSPORTATION COSTS | Lowest cost satellites: simple and light. Lowest cost transmission for each satellite | Moderate cost for each satellite. Moderate transportation cost for each satellite. | Largest, most complex and costliest satellites. Highest transportation cost for each satellite |
| PROJECTED SATELLITE LIFETIME | Shortest satellite lifetime, typically 5 year life | Long satellite lifetime, typically 10 to 15 years | Long satellite lifetime, typically 10 to 15 years |
| **COST AND COMPLEXITY ASSESSMENTS FOR THE OVERALL SYSTEM** | | | |
| ARE HANDOFFS AND/OR CROSSLINKS REQUIRED? | Frequent handoffs and/or crosslinks required | Usually no handoffs or crosslinks required | No handoffs or crosslinks required |
| COST AND COMPLEXITY ESTIMATES FOR THE GROUND CONTROL SEGMENT | Usually the most complex and costly ground control links | Relatively low-cost ground control segment | Relatively low-cost for ground control segment |
| COST AND WEIGHT OF THE PERSONAL COMMUNICATORS | Moderately costly, but highest weight communicators | Moderately costly communicators of moderate weight | Inexpensive but heavy communicators |
| CAN INCREMENTAL STARTUP COVERAGE BE ACHIEVED? | Incremental startup coverage usually not practical | Incremental startup coverage can be practical | Incremental startup coverage very practical |

# Mobile Communication Satellites at Geosync

*You are imposing on us. Do you imagine we are to be fooled by a ventriloquist?*

PROFESSOR BOULLARD
*French scientist who seized Thomas Edison by the throat during a public demonstration of the phonograph, 1915*

Western settlers pondering the awesome power of bright, shining telegraph wires strung across the American frontier dared to dream big dreams. But, even the most imaginative among them would never have dreamed that ordinary Americans would someday be chatting with friends and neighbors with direct relays through orbiting satellites. Those who sailed around Cape Horn severed all family connections for 8 full months. When they arrived on western shores, some were greeted with the news that they had unknowingly become proud parents while in transit on southern seas.

Today's urbanites are superbly connected by cellular phones, so they can stay in constant contact with nearly everyone. But even today millions of Americans who live in thinly inhabited sections of our country do not find it quite so easy to keep in touch. Vast stretches of the North American continent are still not served by cellular telephone. Many remote areas do not, in fact, have telephone service at all.

Fortunately, mobile communication satellites positioned in low-altitude, medium-altitude, and geosynchronous orbits will soon go aloft to fill in the missing gaps on the American continent—and throughout the rest of the world. The beneficial results we can expect from geosynchronous mobile communication satellites are summarized in this chapter. The next three chapters consider mobile communication constellations launched toward lower-altitude destinations.

## American Mobile Satellite's Geosynchronous Constellation

By 1995, with 18 million subscribers hooked together by cellular telephone, industry revenues are expected to top $15 billion per annum with 29 percent of carrier profits coming in from "roaming." A roaming user sends telephone messages through a distant cellular system to which he or she is not a direct subscriber. Roaming messages typically cost about $1.00 per minute, a rate at which some mobile communication satellites can be fully competitive.

Thousands of base stations have been installed, but even so 25 percent of the continental United States and 14 million Americans are still not served by any cellular system. However, with a mobile car phone connected to a brick-sized amplifier installed in the trunk of the vehicle, they can be covered by both satellite relays and their favorite terrestrial cellular system—if one is locally available!

In 1987 the FCC granted the American Mobile Satellite Corporation (AMSC) permission to provide mobile voice, fax, data distribution, and other related services to the United States and nearby territories. These services are to be initiated with a single mobile satellite (MSAT) positioned along the geosynchronous arc. AMSC's most important target markets include domestic communications and public safety organizations, aircraft-to-land message relay, commercial fishing fleets, and coastal shipping firms. Fixed-site telephone connections will also be provided to areas not now being served by local telephone. Company marketeers foresee profitable markets among the 250,000 general-aviation aircraft operating in the United States and among our country's 550,000 pleasure boats measuring at least 28 ft.

In addition, AMSC's specialists are targeting trucking firms and big government agencies in separate marketing campaigns. Private network users are also being encouraged to set up their own independent communication hubs equipped with large parabolic antennas. At least two big trucking firms have already signed on: TMC Transportation and Southeastern Freight Lines. Their drivers currently operate 500 and 1200 big rig 18-wheelers, respectively.

So far, AMSC of Reston, Virginia has been granted authorization for barely enough bandwidth for a single satellite designed to provide 2000 simultaneous duplex voice channels. So a scant 2 percent market penetration into cellular telephone would probably saturate their spaceborne system. In late October of 1994, AMSC will be testing the market with their first satellite. If it is successful, they will probably attempt to obtain larger bandwidth allocations so they can build and launch additional satellites.

AMSC is a complicated international consortium, but 90 percent of it is owned by five large entities: McCaw Cellular Communications (now with AT&T), Hughes Communications, Inc. (now owned by General Motors), Mtel Corporation, Singapore Telecom, and public investors. The AMSC mobile communications units, which will weigh about 10 lb each, are being manufactured by Westinghouse in Baltimore, Maryland, and Mitsubishi-Diamond Tel (with offices in Chicago, Illinois). AMSC's mobile communication satellite is being designed and produced by Hughes in El Segundo, California. Launch services will be provided by General Dynamics in San Diego aboard their Atlas IIA booster rocket.

Each MSAT telephone is a dual-capability unit that operates through both satellite relays and ground-based cellular switches. The system's computers automatically seek a ground-based cellular transmitter to relay each incoming call. If cellular is not available, the call is processed by satellite. The dual-purpose communicators are expected to retail for about $2000 each. Preliminary mobile service rates have been set at $1.00 per minute for commercial users, equipped with high-gain antennas, at $1.45 for the standard service, and $2.50 per minute for aeronautical users. Fixed-site installations, including rural telephone users, are to be charged $0.60 for each minute.

### Antenna Coverage Patterns

American Mobile Satellite Corporation's first geosynchronous satellite will be installed along the geosynchronous arc at 101 degrees west longitude just south of Houston, Texas. The satellite's L-band data communication system will relay digital TDMA messages down toward six spot-beam regions on and around the United States. The frame-length of each burst from the TDMA transmission system is 0.12 second with six messages per frame, one for each of the six satellite spot-beams. The voice messages, by contrast, employ FDMA modulation techniques.

The sketches in Fig. 7.1 represent the coverage patterns of the L-band spot-beams coming down from the satellite. Four of the beams are devoted to the continental United States with a more or less direct correspondence with our country's four time zones: Eastern, Central, Mountain, and Western. The fifth spot-beam stretches across the bottom of the country to provide coverage for Mexico and Puerto Rico and the Caribbean. The sixth beam is split into two separate parts to cover Alaska and Hawaii. In addition to covering the specific land masses to which they are assigned, the various spot-beams, acting together, will also provide coverage out to the 200-nautical mi maritime limit along the continental shelf.

### Designing the Satellite

The MSAT satellite, which weighs 6300 lb, is almost 70 ft long. As Fig. 7.2 indicates, it is equipped with two L-band mesh antennas that are designed to unfurl in space before they begin to handle bent-pipe message relays. The two identical antennas have elliptical cross sections measuring 19.7 by 16.4 ft in their major dimensions.

One of the two L-band antennas is rigged for receive-only, the other operates in the transmit-only mode for message relay. Both antennas, which are mounted on opposite sides of the spacecraft for frequency isolation, operate with right-hand circular polarized carrier waves. The 2.5-ft Ku-band antenna transmits and receives all necessary digital control-segment messages. The carrier waves it transmits and receives are linearly polarized.

An Atlas IIA booster rocket hurls the 3400-lb spacecraft onto an elliptical transfer orbit, with apogee tangent to the geosynchronous altitude. Liquid kick-stage motors circularize the satellite's orbit at geosync. Hypergolic pro-

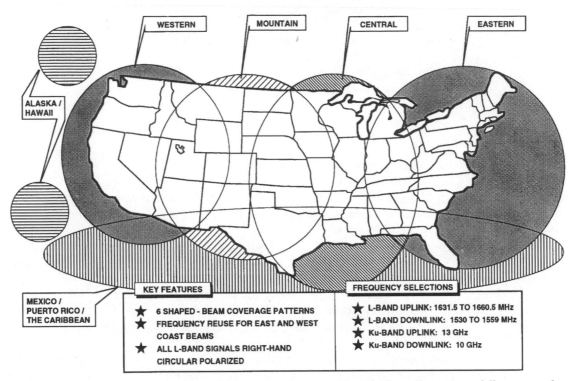

**L-BAND ANTENNA COVERAGE PATTERNS FOR AMSC'S MOBILE COMMUNICATION SATELLITE**

WESTERN    MOUNTAIN    CENTRAL    EASTERN

ALASKA / HAWAII

MEXICO / PUERTO RICO / THE CARIBBEAN

**KEY FEATURES**

★ 6 SHAPED - BEAM COVERAGE PATTERNS
★ FREQUENCY REUSE FOR EAST AND WEST COAST BEAMS
★ ALL L-BAND SIGNALS RIGHT-HAND CIRCULAR POLARIZED

**FREQUENCY SELECTIONS**

★ L-BAND UPLINK: 1631.5 TO 1660.5 MHz
★ L-BAND DOWNLINK:  1530 TO 1559 MHz
★ Ku-BAND UPLINK: 13 GHz
★ Ku-BAND DOWNLINK:  10 GHz

**Figure 7.1**   Six thin spot-beams radiating down toward earth from the MSAT satellite are carefully contoured to provide coverage for the United States and surrounding terrain. Four of the spot-beams correspond roughly to the four time zones stretching across the lower 48 states (excluding Alaska and Hawaii). The fifth beam—which is purposely split into two parts—provides coverage for Alaska and Hawaii. The sixth beam covers large portions of Mexico and Puerto Rico and the Caribbean. Frequency reuse is accomplished between the beams covering the east coast and the west coast of the United States. The other beams are too close together to allow effective frequency reuse.

pellants (monomethyl hydrazine and oxides of nitrogen) fuel the kick-stage motors during on-orbit maneuvers. The two fluids in a hypergolic propellant combination automatically ignite on contact. Hypergolic propellants from the same two tanks also power the satellite's altitude control thrusters.

The initial AMSC satellite will be positioned at 101 degrees west longitude directly south of Houston, Texas, and Topeka, Kansas. East-west station-keeping maneuvers will maintain the satellite's longitudinal position to within 0.1 degree of its assigned orbital slot. Mission requirements also dictate the orbital inclination tolerances which are to be maintained to within 0.1 degree.

Electrical power supplies totaling 3000 watts are generated by the silicon solar arrays mounted on the two opposite sides of the spacecraft. Most of the electrical power is used to drive the various spacecraft subsystems, but a small part of it is stored in the single 123-ampere-hour nickel-hydrogen storage bat-

# ON-ORBIT CONFIGURATION FOR THE MSAT
# GEOSYNCHRONOUS SATELLITE

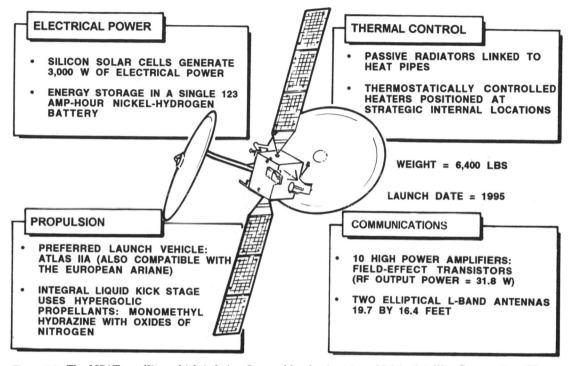

**ELECTRICAL POWER**

- SILICON SOLAR CELLS GENERATE 3,000 W OF ELECTRICAL POWER

- ENERGY STORAGE IN A SINGLE 123 AMP-HOUR NICKEL-HYDROGEN BATTERY

**THERMAL CONTROL**

- PASSIVE RADIATORS LINKED TO HEAT PIPES

- THERMOSTATICALLY CONTROLLED HEATERS POSITIONED AT STRATEGIC INTERNAL LOCATIONS

WEIGHT = 6,400 LBS

LAUNCH DATE = 1995

**PROPULSION**

- PREFERRED LAUNCH VEHICLE: ATLAS IIA (ALSO COMPATIBLE WITH THE EUROPEAN ARIANE)

- INTEGRAL LIQUID KICK STAGE USES HYPERGOLIC PROPELLANTS: MONOMETHYL HYDRAZINE WITH OXIDES OF NITROGEN

**COMMUNICATIONS**

- 10 HIGH POWER AMPLIFIERS: FIELD-EFFECT TRANSISTORS (RF OUTPUT POWER = 31.8 W)

- TWO ELLIPTICAL L-BAND ANTENNAS 19.7 BY 16.4 FEET

**Figure 7.2** The MSAT satellite, which is being financed by the American Mobile Satellite Corporation of Reston, Virginia, will be launched into a geosynchronous orbital slot directly south of the continental United States. Once it reaches its orbital destination, MSAT will furnish mobile communication services to several thousand users equipped with mobile car phones, transportable phones, and telephones mounted at fixed sites, some of which will be powered by flat-panel solar arrays. MSAT will also relay fax messages and digital data using circuit switching and packet switching technology.

tery. The electrical power stored in the battery is used to operate the onboard subsystems when the spacecraft is in the earth's shadow, and, to some extent, during peak-load intervals.

Eight high-power field-effects transistors each produce 31.8 watts of RF output power for each of the two L-band amplifiers. Each satellite is rigged with two separate repeaters and two spare field-effects transistors. The forward repeater receives Ku-band uplinks from the earth-station hubs and relays them at L-band to the mobile telephones. The reverse repeater receives L-band uplink from the mobile telephones and transmits them on the Ku-band to the earth-station hubs.

Three-axis stabilization is provided to the spacecraft using a single, heavy momentum wheel mounted on a swiveling platform. By modifying the platform's orientation and the rotation rate of the momentum wheel, the control system can

---

# A SPACE-AGE WORKHORSE NICKNAMED PERCHERON

It is happening on a tiny island off the southern coast of Texas, an island populated by lethargic cattle and beefalo grazing contentedly on thick, mosquito-infested pastures. In hard hats and swimming trunks, they stride by in animated groups: young, dedicated engineers with dreams as big as Texas ranch country, ready, willing, and able to launch America's first privately financed booster upward toward the space frontier.

They peer out across the green grassy meadows at their strange space-age contraption from their bargain-basement "mission control," an olive drab metal shed hidden behind a protective wall of overstuffed sandbags. It sits all alone on a 12-in. slab of concrete, a 36-ft needle-nosed cylinder similar to the Redstone rocket that 20 years earlier lifted Alan Shepard into space.

They call it their rocket *Percheron,* because, like that sturdy French workhorse, its strength and dependability will weigh heavily on the minds of the well-heeled investors who must join them if they are to make a business of blasting payloads into space.

"Everyone laughed when I first told them about launching a rocket," says chief financier David Hannah, Jr. "Now I think people are starting to believe this dream can come true." Hopes are definitely soaring on Matagorda Island, and everyone who has come to sweat it out with them in that burning Texas sun is hoping, too, even the hard-nosed newspaper reporters shifting from side to side on the green bleachers they have erected for today's demonstration. But, unfortunately, this day's scheduled launch is not to be. A glitch in the ignition system forces a postponement, so the big, disappointed bird must sit silently on its pad for at least another week. Once that difficulty has been ironed out, another countdown will begin. This time, when a tense finger presses the "launch" button, the *Percheron* blows up. However, the following year after fresh infusions of capital—$6 million from 57 investors—another try. And this time *Percheron* finally does fly!

---

create precisely directed action-reaction forces to cause the spacecraft to maintain the proper earth-seeking orientation. A "staring" earth sensor provides all necessary feedback control measurements to the satellite control computer, thus allowing it to fine-tune its attitude control commands.

The MSAT spacecraft maintains its internal temperatures within narrow limits using both passive and active thermal control techniques. Passive radiators mounted on the two opposite sides of the spacecraft, for instance, dissipate excess heat into the blackness of space. Heat pipes carry unwanted heat from the interior of the spacecraft to the external radiators. Electrical resistance heaters help maintain a nearly constant temperature for the propulsion system and the vehicle's electrical power-generating devices. Thermostats rigged with spiral-wound bimetallic strips automatically activate and deactivate the electrical resistance heaters. Under thermal loads, the two metals comprising the bimetallic sandwich expand and contract differentially to loosen or tighten the spring's coil, thus controlling the current feeding the electrical resistance heater.

## Mobile Telephone Characteristics

"According to Frost and Sullivan International, 1996 terminal sales for mobile satellite communications are expected to increase to $1 billion, with annual service revenues surpassing...$472 million," says Patrick W. Baranowsky II of

Westinghouse Electric in Baltimore, Maryland. Westinghouse engineers design and build the Series 1000 mobile phones capable of sending and receiving digital pulse trains to and from AMSC's mobile communication satellite.

Inmarsat's least expensive mobile satellite terminal retails for $15,000 with a $5.50 per minute phone rate. But the Westinghouse Series 1000 is being designed to sell for about $2000, and it will be served by the MSAT satellite at $1.45 for each minute of space-bound conversation.

The bandwidth presently assigned to AMSC can support 2000 full-duplex voice circuits. With reasonable usage rates, this allocated bandwidth can support a few hundred thousand mobile subscribers. Future subscribers could total 2 million or more if all future bandwidth applications are approved and all the presently planned satellites are launched into space.

The transceiver subsystem installed in each mobile terminal consists of a transmitter and a receiver. The transmitter accepts voice, data, and fax together with signaling pulses and converts the analog or digital information they contain into modulated L-band RF signals. The transmitter includes a high-power amplifier and upconverter, and the associated frequency synthesizer. The transmitter also includes modulators and electronic modules devoted to handling voice, fax, and data encoding. The receiver is rigged to accept RF signals from the antenna subsystem and convert their modulations into voice, digital data, fax, or signaling pulses, as required.

Some mobile terminals will also accept precise positioning information from radionavigation systems such as the Loran C or the Navstar GPS. This information may be used to steer the antenna beams toward the satellites or it can be employed for user-desired applications, such as periodic reporting of a vehicle's position and velocity to its parent base station.

Every voice message is digitally scrambled using a prearranged privacy code. The scrambling function is automatic and transparent to the user. Echo cancellers are also activated for all voice-mode connections, but they are bypassed for standard digital data and facsimile transmissions. A special codec multiband algorithm squeezes extra voice information onto a smaller bandwidth. An identical codec routine is being used in connection with the Inmarsat-M digital voice-communication system.

Several different types of antennas are available for use in connection with the MSAT Series 1000 mobile phones. For most land vehicle applications, the antenna of choice will likely be a phased-array flat plate 1 ft in diameter. A mechanical antenna rigged with automatic swiveling capabilities will also be able to produce effective results. The 6-second acquisition time for this unit is quite a bit longer than the acquisition interval for the phased-array antenna, but the mechanical antenna can maintain extremely precise satellite tracking when it is coupled with feedback control devices capable of determining the satellite's current angular location. Signal-strength meters and accelerometers coupled with mechanical or solid-state gyroscopes can provide the required angular-orientation information. A much simpler but smaller omnidirectional mast antenna may find a few niche-market applications. However, studies show that, for effective operation, the mast will have to be at least 3 ft high.

### Facing the Competition

So far AMSC marketeers have negotiated 3-year agreements with at least 150 cellular telephone carriers who will be distributing their services, mostly in connection with cellular telephone. Todd Wolfenbarger, a representative from McCaw Cellular Communications, gets a little ruffled when Inmarsat officials and other marketplace competitors complain about the monopoly AMSC has been granted by the FCC.

That monopoly, he points out, limits their coverage to specific land masses and offshore regions near the continental United States. Moreover, in his view, the monopoly the company has been granted does not make AMSC competitive with presently planned low-altitude voice messaging constellations (to be discussed in Chapter 9), which are being designed to serve hand-held communicators on a global basis.

"Our services will not support a hand-held phone in our first generation," Wolfenbarger points out. "It will support a transportable phone weighing less than 10 pounds with a dual-mode capability to use with both cellular and satellite systems."

### Proposed Constellation Expansions

AMSC's single-satellite mobile communication system is projected to cost $605 million by the time it is fully installed in space. Most of the necessary money has already been raised, concluding with a $236 million public stock offering for 11.5 million shares of company stock filed with the Securities and Exchange Commission in November of 1993. In parallel with AMSC, Telesat Mobile, Inc. of Ottawa, Canada had planned to launch a duplicate MSAT satellite to be positioned at 106.5 degrees west longitude directly south of Denver, Colorado and Moosejaw, Saskatchewan.

Unfortunately, their ambitious plans hit a major snag when, in April of 1993, Telesat Mobile, Inc. was forced to file for bankruptcy protection from its creditors. Telesat officials had been struggling financially for months after $213 million was suddenly withdrawn by financial backers. At last report Bell Canada had stepped in to take over for Telesat Mobile with sufficient funds to launch the Canadian version of MSAT into space.

AMSC has been authorized to launch a total of three geosynchronous satellites. But, before Telesat Mobile collapsed, company officials had, at times, talked about a total constellation of eight. Six of them would be owned and operated by AMSC. The other two were to be owned and operated by Telesat Mobile. Planned orbital slots were located at 62, 101, 106.5, and 139 degrees west longitude, with two MSAT satellites occupying each orbital slot.

### Inmarsat's Project 21

The architects of Inmarsat's Project 21 (which is sometimes called Inmarsat-P) may end up using geosynchronous satellites to serve the needs of thousands of mobile users. Project sponsors initially evaluated three different satellite constellations each capable of delivering global voice connections for about $1.00 a minute.

The low-altitude constellation they had evaluated was to be populated by 54 mobile communication satellites orbiting in 1125-nautical mi circular orbits. Once their preliminary studies were complete, company officials rejected the low-altitude option because they believed it was an excessively costly solution to the communication problem they were attempting to solve. The two remaining options, which are slated to employ medium-altitude and geosynchronous satellites, are still being evaluated by various teams of engineers and marketeers.

The medium-altitude option is baselined as a supplement to Inmarsat's conventional four-satellite geosynchronous constellation. If this option is selected, it will consist of 9 to 15 satellites in circular orbits with altitudes lying somewhere between 5400 and 8100 nautical mi.

Inmarsat's geosynchronous constellation would be similar to the one the organization now operates with four geosynchronous satellites positioned at 15.5 and 55 degrees west longitude and 64.5 and 178 degrees east longitude. The new mobile communication satellites would have enhanced design features with more powerful RF transmitters, more efficient digital modulation techniques, and enlarged spacecraft antennas. Inmarsat officials supported by a global team of space-systems engineers are still trying to pin down the best available mobile satellite constellation. Some of the participants believe that Inmarsat—or a portion of it—should be formed into a private, profit-making entity to make it more flexible and competitive.

## Tomorrow's Global Family of Mobile Comsats

Large numbers of mobile communication satellites are being planned, and in some cases, launched by various teams of aerospace specialists in many different parts of the world. Unfortunately, the regionally oriented backers of these various mobile communication systems have not been coordinating their efforts, so their technical design features and their modulation formats will all be different. These differences will make it even more difficult for satellite operators to coordinate future use of the frequency spectrum and to develop standardized receivers to achieve the economies of scale.

In November 1994, Mexico's communication experts plan to launch their second Solidaridad satellite with a mission for providing mobile communication mainly to trucking fleets. Despite intensive efforts in a number of meetings, Mexican officials have been unable to coordinate their choice of frequencies and bandwidths with their counterparts in the United States, Canada, and other nearby countries.

In 1994, Optus Communications of Sydney, Australia started providing mobile telephone services using digital L-band signals relayed through their Optus B-1 satellite, which went into orbit in September 1992. Services are provided within Australia and in surrounding waters using briefcase-sized satellite telephones built by Westinghouse. The Optus B-3 satellite, which recently flew up to the geosynchronous altitude aboard a Chinese Long March rocket in mid-1994, will also include mobile communication relay links.

Eutelsat, which is to be launched by European Telecommunications in Paris, will allow mobile users to exchange short, digital telegram messages on the fly. European space agency experts are, in addition, planning to conduct a number of mobile communication experiments in connection with Italy's Italsat now slated for launch some time in 1995.

Also in 1995, aerospace experts from India should be ready to launch their Insat IC communication satellite, which will carry a payload capable of providing the first mobile communication services to the Indian subcontinent. Japanese space scientists are already operating two satellites that handle mobile communication services for Japan's trucking industry and other related users. But those Japanese satellites employ older and less effective analog modulation formats. Nevertheless, two more of them are scheduled to go into service in 1994.

Later, the Japanese ETS-6 S-band payload will provide spot-beam coverage for Japan's mobile phone users equipped with small hand-held terminals and their new N-Star satellite will provide substantially improved services. A separate group of Japanese entrepreneurs has also invested millions of dollars in Motorola's ambitious low-altitude Iridium constellation.

## Elliptical Orbit Constellations

Today's geosynchronous communication satellites are all being positioned in circular orbits with orbital inclination angles of, at most, only a few degrees. Such a satellite stays close to the center of its assigned orbital slot, so antennas on the ground can access it by pointing toward a specific point in the sky. Circular geosynchronous orbits have served the developed world extremely well for many years, but important benefits can be achieved by launching communication satellites into inclined elliptical orbits.

### Molniya Constellations Perfected by Soviet Rocketeers

Perhaps the biggest disadvantage of a geosynchronous satellite is that it does not provide adequate coverage at extreme northern and southern latitudes. At those latitude extremes, the electromagnetic waves coming down from a geosynchronous satellite pass very close to the horizon where blockages, atmospheric distortion, and rain attenuation can be frequent and severe.

Because so many citizens of the former Soviet Union live at high northern latitudes, early Soviet engineers tended to shun geosynchronous satellites. Instead, they have launched most of their communication satellites into inclined elliptical Molniya (Mol-nee-ah) orbits.

Figure 7.3 highlights the orbital geometry and the hardware configuration for a typical Molniya satellite. Notice that the satellite is launched into a 12-hour orbit with a 63.4 degree orbital inclination. Notice also, that its apogee has been purposely positioned in the northern hemisphere high above the former Soviet Union. The perigee altitude of a typical Molniya orbit is 500 nauti-

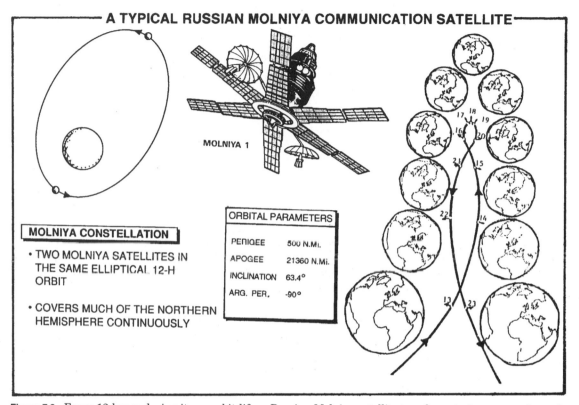

**Figure 7.3**  Every 12 hours during its on-orbit life, a Russian Molniya satellite travels up to its apogee above the northern hemisphere where it acts as a communications relay station for the Russian people who live at high northern latitudes. Each Molniya satellite is launched into space with a 63.4 degree orbital inclination, so its apogee will not be perturbed by the earth's equatorial bulge to drive it southward away from its northernmost latitude. When a Molniya satellite is near apogee over former Soviet territory, it moves very slowly across the sky. Consequently, during that interval it behaves much like a geosynchronous satellite that provides superb communications coverage at extremely high latitudes.

cal mi and its apogee is 21,300 nautical mi, just a bit higher than the geosynchronous altitude. When it moves up toward apogee, a Molniya satellite travels very slowly across the northern sky. Thus, it behaves much like a geosynchronous satellite at an extremely high latitude. Only the most gentle ground antenna movements are necessary to keep it in view.

Molniya's 63.4 degree inclination angle was specifically chosen because any satellite with such an inclination has a zero rotation rate about the line of apsides joining its apogee with its perigee. Therefore its apogee always remains at the northernmost latitude of the satellite's orbit (63.4 degrees). The "double" figure eight ground trace highlighted in Fig. 7.3 results from the motion of the Molniya satellite as it swings up toward apogee combined with the earth's constant 15 degree per hour rotation out from under the satellite.

A four-satellite Molniya constellation properly positioned in space can provide continuous 24-hour per day coverage for the entire northern hemisphere. Such coverage capabilities may seem rather impressive. But actually, as the

American researcher John Draim showed in 1987, a properly constructed four-satellite constellation can provide continuous coverage for both the northern and the southern hemispheres.

## John Draim's Four-Satellite Constellation

In 1987, after 7 worrisome years of pondering the problem, drawing numerous sketches, and making careful computer simulations, aerospace engineer John Draim developed a four-satellite constellation that provides continuous coverage for the entire surface of the earth.

Until that time, most mission analysis specialists were convinced that five satellites was the smallest number that could do the job. In the 1960s, an English researcher named John Walker demonstrated that five satellites launched into properly chosen 6200-nautical mi circular orbits or higher inclined about 55 degrees with respect to the equator could provide continuous global coverage. Later experts attached the name "rosettes" to the Walker constellations because, when they are viewed from a vantage point above the North Pole, they resemble roses in full bloom.

John Draim's cleverly designed constellation reduces the number of satellites needed for continuous global coverage to only four. Those four satellites are launched into 31 degree inclined elliptical orbits with orbital periods of 26.5 hours or greater. A few years ago John Draim came to one of my orbital mechanics short courses in Washington, D.C. Later, while we were chatting in the hotel lobby, I told him that, when I had constructed the Navstar's 24-satellite constellation, I temporarily froze the rotation of the earth. This, I went on to explain, helped simplify the analysis procedures. "That's interesting," John Draim replied, "I let the earth rotate every 26.5 hours in my preliminary studies—for exactly the same reason."

John Draim has developed a modified version of his continuous-coverage constellation that employs (slightly lower) 24-hour elliptical orbits. At that lower average altitude, the constellation does not quite provide continuous coverage for the earth, but, it comes very close. On the average, everywhere in the world, at least one satellite from his constellation is in view 99.996 percent of the time!

The sketches in Fig. 7.4 focus on some of the more important design features of John Draim's four-satellite, 24-hour constellation for which he, incidentally, holds an official United States patent. The four satellites, in pairs, trace out the two egg-shaped ground traces 180 degrees apart along the equator. By the way, at all times, all four of the Draim satellites have continuous intervisibility; each one of them can always see the other three.

Because the satellites in the Draim constellation are positioned in 31 degree orbits, they can be launched almost due east out of Cape Kennedy, Florida, with no requirement for expensive on-orbit plane-change maneuvers. Such an inclined orbit requires less propulsive energy so the satellite can be 25 percent heavier than a comparable geosynchronous satellite launched by the same rocket.

So far, John Draim's four-satellite constellation has not been used for any practical mission flown into space. But the architecture he perfected has a

# JOHN DRAIM'S 4-SATELLITE ELLIPTICAL-ORBIT CONSTELLATION

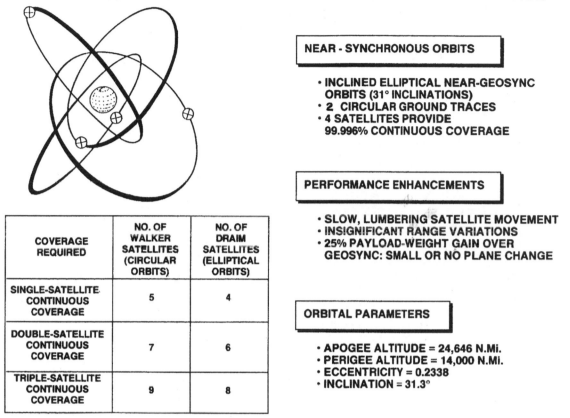

**NEAR - SYNCHRONOUS ORBITS**

- INCLINED ELLIPTICAL NEAR-GEOSYNC
  ORBITS (31° INCLINATIONS)
- 2 CIRCULAR GROUND TRACES
- 4 SATELLITES PROVIDE
  99.996% CONTINUOUS COVERAGE

**PERFORMANCE ENHANCEMENTS**

- SLOW, LUMBERING SATELLITE MOVEMENT
- INSIGNIFICANT RANGE VARIATIONS
- 25% PAYLOAD-WEIGHT GAIN OVER
  GEOSYNC: SMALL OR NO PLANE CHANGE

**ORBITAL PARAMETERS**

- APOGEE ALTITUDE = 24,646 N.Mi.
- PERIGEE ALTITUDE = 14,000 N.Mi.
- ECCENTRICITY = 0.2338
- INCLINATION = 31.3°

| COVERAGE REQUIRED | NO. OF WALKER SATELLITES (CIRCULAR ORBITS) | NO. OF DRAIM SATELLITES (ELLIPTICAL ORBITS) |
|---|---|---|
| SINGLE-SATELLITE CONTINUOUS COVERAGE | 5 | 4 |
| DOUBLE-SATELLITE CONTINUOUS COVERAGE | 7 | 6 |
| TRIPLE-SATELLITE CONTINUOUS COVERAGE | 9 | 8 |

**Figure 7.4** In 1987, the American engineer John Draim demonstrated that four satellites properly positioned in 26.5-hour inclined elliptical orbits can provide continuous coverage for the entire surface of the globe. Another version of his four-satellite constellation, which is sketched in this figure, relies on four 24-hour satellites to provide global coverage 99.996 percent of the time. Although the Draim satellites travel slowly around their circular ground traces, they can be launched into orbit with much smaller propellant supplies than conventional geosynchronous satellites. This is true because the 31-degree Draim constellations do not require expensive on-orbit plane-change maneuvers. (*Courtesy of* Guidance and Control Journal, *1991.*)

number of theoretical advantages. So someday soon his unique constellation of satellites may be adopted by one or more profit-minded entrepreneurs.

## Russia's Humongous Orbital Antenna Farm
## Built on a Massive Scale

Soviet engineers launched their first geosynchronous communication satellite in 1974, about 10 years after America's rocketeers began sending theirs into space. In the meantime, however, Soviet and Russian scientists have orbited several different families of geosynchronous satellites including Raduga, Ekran, Gorizont, and Luch.

## WATCHING AN APPLE FALLING FROM A TREE

In 1665, Isaac Newton left Cambridge University and returned to his hometown of Woolsthorpe to escape the worst ravages of the Black Plague. Safely back among familiar surroundings, he made landmark discoveries that have provided us with precisely the keys we needed to conquer space.

Although the young Newton had reportedly been a mediocre student in the early grades, his powerful intelligence asserted itself even before he reached his teenage years. When he was still a tow-headed youngster, for instance, he managed to construct a charming little windmill backed up by one mouse-power so it could go on turning when the wind refused to blow. Later, he made a paper kite rigged to carry a small lantern high above the British countryside. The people of Woolsthorpe had never before seen flickering lights floating across the nighttime sky, so the young Isaac may have been responsible for some of the earliest sightings of UFOs.

At the age of 23, while relaxing on his mother's farm, Isaac Newton, by his own account, saw an apple falling from a tree. That simple incident caused him to wonder why apples always tumble down. That apple tumbled down toward the ground while the pale August moon continued to sail contentedly overhead. Soon he theorized that the force of gravity tugging on apple and moon falls off systematically with increasing altitude in the same way a light beam dissipates as we move farther away from its source. Double the distance and its intensity falls off by a factor of 4.

Thus, by Newton's reckoning, the force of gravity pulling on the moon should be about 1/3000th as strong as the gravity we experience at the surface of the earth. In 1 minute, he soon calculated, a falling apple would be pulled downward about 10 mi, but the moon would fall toward the earth only about 16 ft. During that same 1-minute interval, the moon's orbital velocity also carried it sideways 38 mi. Consequently, its horizontal and vertical motion combine to bring it back onto the same gently curving circular path over and over again.

Isaac Newton figured out how gravity works because of a fortunate encounter with his mother's favorite apple tree. Armed with only his inverse square law of gravitation, three deceptively simple laws of motion, and one of the most powerful intellects that ever pondered anything, Newton quietly set about to unravel the hidden secrets of the universe.

In 1991, Russia's aerospace experts proposed an enormous orbital antenna farm called Globis to be launched directly up toward the geosynchronous altitude aboard their equally enormous SL-17 Energia booster. The projected on-orbit weight of Globis is 42,000 lb and it generates 15,000 watts of continuous electrical power to operate its electromagnetic transmitters and other onboard subsystems. Globis is being designed to provide two dozen channels of color television, plus 300,000 full-duplex voice channels for fixed ground stations, and 1400 circuits for mobile pocket telephones. Transmission frequencies are being planned for several bands in the electromagnetic frequency spectrum ranging from 1 to 18 GHz.

Figure 7.5 highlights some of the key features of the Globis satellite, which is being designed with at least 30 antennas and two gigantic solar arrays. Its on-orbit design life is 10 full years, so it will have to be built with abundant numbers of redundant subsystems.

Now that the economy of the former Soviet Union has deteriorated to such a sorry state, the Globis orbital antenna farm may never be constructed. But, nevertheless, it is a rather intriguing example of entrepreneurial imagineering, particularly when compared with its American counterpart, which is sketched in Fig. 2.12.

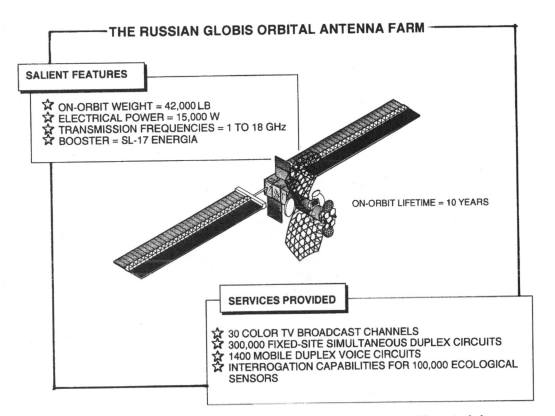

THE RUSSIAN GLOBIS ORBITAL ANTENNA FARM

**SALIENT FEATURES**

☆ ON-ORBIT WEIGHT = 42,000 LB
☆ ELECTRICAL POWER = 15,000 W
☆ TRANSMISSION FREQUENCIES = 1 TO 18 GHz
☆ BOOSTER = SL-17 ENERGIA

ON-ORBIT LIFETIME = 10 YEARS

**SERVICES PROVIDED**

☆ 30 COLOR TV BROADCAST CHANNELS
☆ 300,000 FIXED-SITE SIMULTANEOUS DUPLEX CIRCUITS
☆ 1400 MOBILE DUPLEX VOICE CIRCUITS
☆ INTERROGATION CAPABILITIES FOR 100,000 ECOLOGICAL
  SENSORS

**Figure 7.5**  Russia's orbital antenna farm called Globis is being designed for launch by a single humongous Energia booster directly toward a 19,300-nautical mi geosynchronous orbit. The Globis design weight is 42,000 lb with enough capability for 30 direct-broadcast television channels and 300,000 full-duplex voice channels serving ground-based antennas at fixed sites. The highly ambitious Russian Globis may never be launched into space. But it is an intriguing concept, in part because it so closely resembles its American counterpart developed in 1977 in conjunction with Rockwell International's space industrialization study efforts.

# 8

# Telegram Satellites
# Launched into Low-Altitude Orbits

*With regard to electric light, much has been said for
and against it, but I may say, without fear of
contradiction, that when the Paris exhibition closes,
electric light will close with it, and very little more
will be heard of it.*

PROFESSOR ERAMUS WILSON
*Commenting on the incandescent light bulbs used in
decorating the Paris Exhibition, 1878*

"Can three Harvard MBAs and a Texas wildcatter take on the aerospace establishment? Tune in for the continuing Adventures of Orbital Sciences." That playful little lead sentence was written by Julius Ellis in *Air and Space* published by the Smithsonian Institution in Washington, D.C. The rousing saga it describes started when the bright young founders of Orbital Sciences Corporation, Bruce Ferguson, David Thompson, and Scott Webster—bubbling with confidence and optimism—formed their fledgling company in the late fall of 1982. In their first successful venture they built and marketed a high-flying upper-stage rocket called the Transfer Orbit Stage.

The first unscheduled meeting between Ferguson, Thompson, and Webster took place in a classroom at Harvard University in 1979 when they all enrolled in a NASA-sponsored university project intended to evaluate the many opportunities for the commercialization of space. Some of their classmates scoffed at the space-age concepts under review, but those three tight friends quickly focused on a near-term opportunity—an upper stage to be released from the cargo bay of NASA's reusable space shuttle.

From that first day forward, the talents of the three young students meshed together like a superbly crafted Swiss watch. Thompson had graduated number one in his class at Massachusetts Institute of Technology (MIT) in aeronautics and astronautics, Webster was a mechanical engineer, and Ferguson had pursued a joint degree in law and business.

David Thompson had been a self-proclaimed space cadet ever since his teenage years when he built a crude 7-ft rocket that hurled squealing monkeys on mile-high parabolas above the fertile farmlands outside Spartanburg, South Carolina. He had always enjoyed pretending he was Christopher Krafft at Mission Control, but he never pondered the possibility of trying to form a space company until he enrolled in that fateful workshop at Harvard University. Nor had he ever considered "holding a board meeting in outer space," a dream he eventually shared with the other two self-styled space cadets.

"Upper stages seemed ideally suited to private investment because of their fairly simple technology, modest development cost, and large customer base," Bruce Ferguson later observed in a magazine interview. "Yet only McDonnell Douglas had taken the initiative to fund its own shuttle upper stage." Lightweight payloads were adequately covered by Boeing Inertial Upper Stage. Consequently, the dynamic trio concluded that a midsized shuttle-compatible rocket might find a comfortable niche smack dab in the middle of the marketplace. Their studies quickly revealed that the new stage could have a number of commercial customers and it was suited to government missions, too.

When they graduated in 1981, Thompson and his colleagues went off in different directions. They probably would have stayed apart if they had not happened to win a $5000 award from the Space Foundation, a nonprofit organization of Houston business executives. The award barely covered their travel expenses to Houston to attend the ceremony, but the honor it bestowed inspired the young entrepreneurs to moonlight for a while to flesh out the details of their fledgling upper-stage rocket.

Later, when they delivered their first appeal for funds to Sam Dunnam, president of the Space Foundation, he directed them to Fred Alcorn, a Houston wildcatter who was ready, willing, and able to engage in high-risk ventures. "I may have been naive," Alcorn recalls. "But the young men's ideas made so much sense, I knew right away I wanted a piece of the action."

On the basis of a proposal sketched on a deli napkin, Alcorn provided them with $30,000—and a promise of a lot more money if and when NASA gave its blessing to the project. Later, Thompson and Webster heard a presentation by a NASA engineer outlining the need for a midsized upper stage similar to the one they already had in mind. The representatives at NASA patiently listened to their presentation, but they were rather noncommittal. Nevertheless, all three of the young engineers quit their jobs and secured another $250,000 from the deep pockets of Fred Alcorn, plus a $2 million line of credit from his bank.

Even so, funds were gradually growing thin until a supportive article published in the *New York Times* led Nathaniel Rothschild, a member of Europe's noted banking family, to form an investor group to pump in another $1.8 million of painfully needed capital reserves. This was followed by a limited partnership good for another $50 million—the largest amount of money ever raised in this manner in the aerospace industry.

When the first Transfer Orbit Stage rolled off the production lines it was scheduled to trace out a graceful trajectory bound for Mars. Unfortunately, the *Challenger* disaster wiped out much of the rest of the little company's potential business. At that fateful meeting in Washington, NASA had pinpointed a need for 10 additional upper stages valued at $30 million each, including the one their representatives eventually agreed to purchase from Fred Ferguson's little company. Still, if Ferguson, Thompson, and Webster were beginning to get discouraged they never seemed to let it show. They confidently began to scan the horizon for other promising opportunities in aerospace. Thompson, Ferguson, and Webster still gaze upward with their eyes tightly focused on the stars. "We can pull off that board meeting in space by the year 2000," Thompson confidently concludes. "I don't know exactly when, but we'll manage somehow."

Following their intermittent success with the Transfer Orbit Stage, the leaders at the Orbital Sciences Corporation devised two additional ways to pull in big bucks from the space frontier: a winged booster rocket called *Pegasus* and a mobile communication constellation to relay digitally encrypted telegram-style messages through 26 Orbcomm Microstar satellites.

The all-solid Pegasus booster will be used to carry the Orbcomm satellites into space with eight Orbcomms flying in a cluster on each mission.

## The Orbcomm Mobile Communication System

Orbcomm's initial constellation will consist of only two satellites launched into polar orbits at an orbital altitude of 424 nautical mi. Such a small constellation will not, of course, provide continuous coverage for any of the users located on the ground. Instead, the satellites will pass over the continental United States eight times every day with a typical pass lasting about 10 minutes.

David Thompson and his professional colleagues at Orbcomm are planning to add 24 additional satellites at that same 424-nautical mi altitude arranged in three orbital rings with 45 degree orbital inclinations. A diagram showing how all four orbital rings are to be positioned with respect to one another is presented in Fig. 8.1, which also highlights the key features of the hand-held communicators that relay telegram-style messages up through the Orbcomm satellites and back down toward the ground below.

Each 12-oz communicator will provide both message relay capabilities and fairly accurate geographic positioning for the user. Positioning solutions are accomplished primarily by means of the Doppler shift variations caused by the systematic movements of the satellites as they sweep across the sky. System designers are anticipating positioning errors of about 1200 ft under routine operating conditions. Any users who require more precise positioning solutions can insert Navstar GPS chips into their digital communicators.

## The Orbcomm Communicators

The Orbcomm communicators will be equipped with 20-in whip antennas and small battery packs capable of generating 5 watts of electrical power. Some of the units will be attached to ordinary car- and truck-mounted radio aerials doing double duty as Orbcomm communication antennas. The communicators are expected to retail for about $100 to $400 depending on their level of engi-

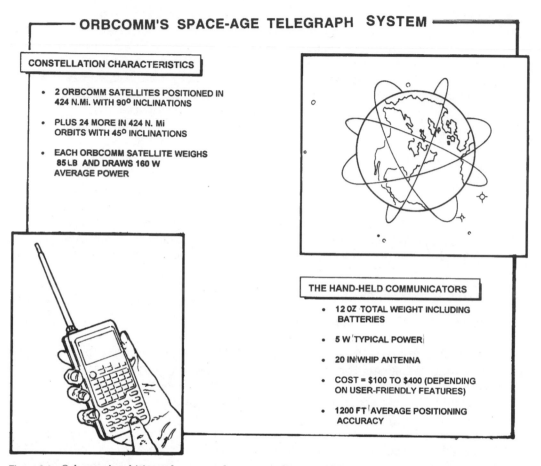

**Figure 8.1**   Orbcomm's orbiting telegram-style system relies on 24 Microstar satellites launched into 45 degree, 424-nautical mi orbits supplemented by two additional polar-orbiting Microstar satellites in the same orbital altitude regime. Each 85-lb satellite generates an average power of 160 watts. The hand-held user sets, which weigh less than 1 lb each, operate at a typical power level of 5 watts. Orbcomm's engineers are still fiddling with various user-set designs, in which they are working toward unit costs in the $100 to $400 range.

neering sophistication. Most of them are being equipped with alphanumeric keypads and viewing screens for digital alphanumeric messaging. Typical messages will cost only a few cents each. The initial communicators are being designed and produced by teams at Matsushita Electric (Panasonic of Japan).

Data rates for the Orbcomm communicators will be 2400 bits/second for the uplinks and 4800 bits/second for the downlinks. The two links operate at 148 to 150.05 MHz and 137 to 138 MHz, respectively. The two-way digital pulse trains can span any desired message length, but typical messages are expected to range from about 6 to 250 alphanumeric characters. Messages in that size range will probably produce the most economical results. Each alphanumeric character to be transmitted will consist of eight adjacent binary digits, so a typical 250-

character message will require about 2000 bits with a corresponding transmission interval of a little less than 1 second.

The data exchange protocols are being structured so they will be compatible with the computer-based E-mail system. The messages can also be routed to or from any suitably equipped fax machine with computerized error detection algorithms ferreting out most transmission errors. All messages are automatically acknowledged by the recipient. By 1998 Orbcomm officials are planning to offer digital communication services at various locations throughout the world.

### Financing the Constellation

In April 1993, Orbital Sciences Corporation officials announced that they had successfully secured the $135 million needed to complete the initial phase of their low-altitude data-relay system. Teleglobe, Inc. of Montreal, Canada agreed to invest $80 million in the project with Orbital Sciences coughing up the other $55 million.

The entire system with all 26 satellites in place is expected to cost a total of $400 to $500 million. Company officials are also reviewing the potential benefits of adding eight more Microstar satellites to provide improved coverage for the system's users and a more robust constellation.

### Building Orbcomm's Microstar Satellites

A sketch highlighting some of the key features of the Orbcomm Microstar satellites is presented in Fig. 8.2. In their stowed configuration the 85-lb satellites are shaped like flat disks so they can be stacked one atop the other. A single Pegasus booster released from the wing of a Lockheed L-1011 should be able to carry eight Microstar satellites into their low-altitude earth orbits.

The satellites, which some observers have likened to "coins on a roll" are only 41 in. in diameter and about 6.5 in. high. They are to be carried to their destination orbits by the three-stage all-solid Pegasus booster. Coiled springs and cold nitrogen gas thrusters will be used to space all eight satellites 45 degrees apart around their orbital ring. When each coin-shaped Microstar satellite reaches its proper orbital destination, its two flat-panel solar arrays will be deployed to generate electrical power (Fig. 8.2) and its 8.5-ft gravity gradient boom will be extended downward toward the center of the earth.

Gravity gradient stabilization has been used in controlling the attitudes of orbiting space vehicles for more than 30 years. It was pioneered by researchers at the Applied Physics Laboratory at Johns Hopkins University 30 mi or so from Baltimore, Maryland. Gravity gradient stabilization is based on the fact that the gravitational force acting on the lower end of an elongated satellite pointing toward the earth is stronger than the gravitational force acting on its upper end.

Thus, if an elongated satellite is oriented so its long axis points toward the center of the earth, it will tend to maintain that stable earth-seeking orienta-

# ORBCOMM'S DISK-SHAPED MICROSTAR SATELLITE

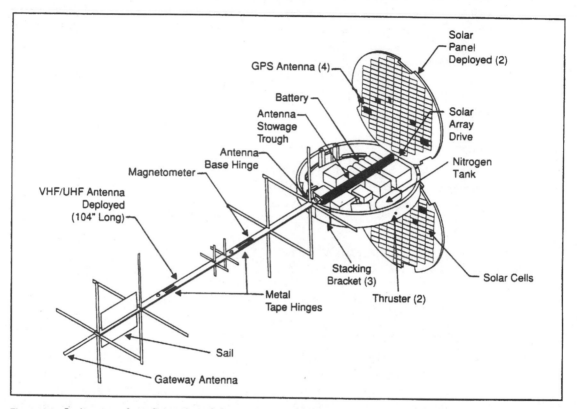

**Figure 8.2**  In its stowed configuration, Orbcomm's disk-shaped Microstar satellite is only 6.5 in. high. Once it reaches its orbital destination, the satellite automatically deploys two pie-plate–shaped solar arrays for electrical power generation and a single 8.5-ft gravity gradient boom to achieve earth-seeking attitude stabilization. For maximum economy, Orbcomm's design specialists are expecting to have each Pegasus booster carry as many as eight Microstar satellites into space.

tion as it coasts around its free-fall orbit. If some random set of perturbations begins to rotate it away from the vertical, the gravity gradient will gently tug it back toward its original earth-seeking orientation.

For a Microstar satellite in a 424-nautical mi orbit, the gravitational force on the end pointing toward the center of the earth is only about 0.00007 percent stronger than the gravity on the upper end. This natural force difference may seem exceedingly small by the standards of everyday life. But in the relatively benign environment of outer space, it is strong enough to maintain a reasonably reliable earth-seeking orientation for elongated satellites in low-altitude earth orbits. Gravity gradient stabilization is a passive attitude control technique that does not require the consumption of precious on-board propellants. It works best when it is used in connection with satellites in low-altitude orbits. At substantially higher altitudes the forces created by the gravity gradient are so small that enormously long, rigid, and delicately balanced booms are required to produce sufficiently large torques to maintain an earth-seeking spacecraft orientation.

The individual satellites in the Orbcomm constellation are relatively small and surprisingly compact. But taken in conglomerate, all 26 of them can provide some rather impressive communication capabilities. As space cadet David Thompson likes to tell visitors to the Orbital Sciences Corporation: "A constellation of 26 [Microstar satellites] will give the system a total of 6 kW of power in 2200 lb of space hardware—or twice the power and less than 25 percent the weight of an Intelsat 6, currently the most powerful geosynchronous telecommunication satellite."

Each Microstar satellite will include 17 separate and distinct data processors together with seven small antennas which, taken together, will allow it to relay some 50,000 digital messages during every 24-hour interval it is in operation. Moreover, Orbital Sciences officials are convinced that "the entire Orbcomm system can handle 5 million messages every day."

The Orbcomm satellites are designed to maintain an earth-seeking orientation to within an accuracy of ±5 degrees. Their average power consumption is 160 watts with nickel-cadmium batteries for energy storage. Each $1.2-million satellite is designed for a 4-year on-orbit mission life.

The company's carefully crafted market projections call for about 300,000 paying customers by late 1996, a number that seems entirely reasonable in view of the explosive growth rates associated with the cellular telephone industry and the early spaceborne messaging systems such as the Airfone System carried aboard commercial jetliners and the OmniTRACS System now serving commercial trucking firms.

"Looking over the next 4 or 5 years I expect Orbcomm to be our single largest business and perhaps approach in total value everything else we are doing," David Thompson confidently predicts. Some industry commentators have noted that, in silhouette, the 85-lb Microstar satellite looks a little like a mechanical, space-age version of Mickey Mouse with a circular spacecraft body and two solar panels positioned on opposite sides like big, floppy mouse ears.

But, otherwise, there is nothing Mickey Mouse about the business revenues the feisty little satellite seems likely to generate. "Orbcomm will be the world's first global personal communication system," Thompson flatly predicts without hesitation. "It will work everywhere: in cities, in rural areas, in mountains, across deserts, and over the oceans."

## Launching the Satellites into Space Aboard the Pegasus Booster

The Pegasus booster will carry Orbcomm's capable little Microstar satellites into orbit packaged in elongated cylindrical eight-packs. The Pegasus itself looks like a 49-ft torpedo slung under its own flat delta wing (Fig. 8.3). Pegasus was designed to be released from the wing of a Lockheed L-1011 (or a Boeing B-52) at high subsonic speeds. A dedicated team of only 35 engineers at the Orbital Sciences Corporation designed and built the sleek little rocket using a scant $45 million in private capital.

## BIRTH PANGS FOR PEGASUS

More than 30 years ago test pilot Scott Crossfield plunged downward through pale blue skies after he was released from the wing of a B-52 as the lone passenger on board the experimental X-15 aircraft. That particular flight ended with a soft landing, but on a later flight Crossfield crunched down so hard that he snapped the X-15 in half. Before the test program for the X-15 was terminated, some imaginative aerospace engineers believed that it should release rocket-powered payloads that would be flown much faster and higher into outer space.

Years later staff members at the Orbital Sciences Corporation were seeking an inexpensive way to launch their network of mobile communication satellites when their vice president of engineering, Antonio Elias, suddenly noticed that his thoughts were racing back and forth between the X-15 slung under the B-52's right wing and an American antisatellite weapon that was ready to be launched from an F-15.

His exciting mental images of those two air-launched rockets hurtling upward toward the fringes of space prompted Elias to sketch his plans for a rather striking project around the edges of a cafeteria napkin. "My sketch showed a rocket made small and inexpensive," Elias later explained to a reporter, "because a carrier aircraft would serve as its first booster stage."

In addition to the obvious benefit of getting started at a higher altitude, the rocket ignites its engines while flying horizontally—the direction it eventually needs to be traveling to gain orbital speed. The aerodynamic stresses it must withstand are also lower at the high-altitude ignition point, and its rocket engines operate more efficiently in the thinner air.

David Thompson and other officials at the Orbital Sciences Corporation were skeptical at first but eventually they came to embrace the idea of a sleek little rocket being dropped into the air from an airplane wing. Their goal turned out to be a booster they could sell for around $10 million, eventually capable of delivering a 900-lb payload into a low-altitude polar orbit. On their first test flight at Monterey, California, their test engineers came reasonably close to reaching their ambitious goal.

Pegasus is not, by any means, the ultimate launch vehicle, but it does represent a genuine attempt to lower the cost of orbiting a satellite to a more affordable level. "I'm impatient with anyone who says there must be a better way of doing something and then doesn't do it," David Thompson quietly concludes.

All three of the solid rocket stages for the Pegasus are constructed from high-technology graphite–epoxy composite materials, but some of the other on-board modules were lifted from the lower end of the technology scale. The flight-control computer, for instance, was borrowed from Israel's main battle tank, and its inertial guidance system was patterned after a Litton-produced system that the Litton engineers had developed for use on the American Navy's Mark 48 torpedo.

On a typical launch, the L-1011 parent craft releases the three-stage all-solid Pegasus at an altitude of about 40,000 ft. At that altitude Pegasus is cruising above 75 percent of the earth's atmosphere and it is traveling just a little below supersonic speed. A flight into space aboard the Pegasus currently costs approximately $10 million with a maximum payload weight of about 900 lb. The avionics system chosen by the Pegasus designers is based on a distributed architecture with a total of 15 separate microprocessors scattered throughout the innards of their sleek little machine. This rather unique architectural approach makes electronic testing quite a bit easier than it would otherwise be. The test engineer merely plugs an IBM PC computer into each subsystem, then types in a few selected commands to see if the proper response is successfully achieved.

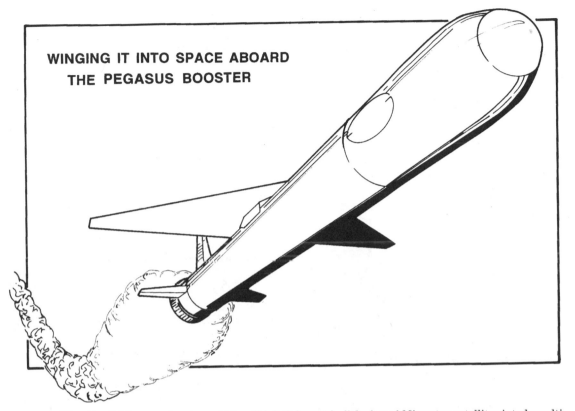

**WINGING IT INTO SPACE ABOARD THE PEGASUS BOOSTER**

**Figure 8.3** The all-solid *Pegasus* booster, which will loft Orbcomm's disk-shaped Microstar satellites into low-altitude orbits, is carried up to its 40,000-ft release altitude under the wing of a modified Lockheed L-1011 commercial jet. When it is released, the stubby wedge-shaped wings attached to the Pegasus booster supply lifting forces to keep the vehicle aloft while its three solid rocket motors hurl the payload up toward orbital speed. Designers of the Pegasus enjoy pointing out that their three-stage Pegasus booster is only about half as heavy as it would have to be if it started off from the ground instead of riding as an aircraft passenger up toward space.

### The Starsys Approach

The Starsys mobile communication system being developed by Starsys Global Positioning, Inc. of Lanham, Maryland is a two-way telegram-style system consisting of 24 store-and-forward satellites to be placed in 60 degree 702-nautical mi orbits. The Starsys satellites will have little or no on-board processing and no data storage capabilities.

The initial Starsys constellation (Fig. 8.4) will consist of two satellites. Assuming that FCC approvals can be obtained, the other 22 are to be deployed as rapidly as possible through the mid-1990s until the constellation is complete. Each Starsys satellite will weigh 220 lb with a design life of 5 years.

Ground-based communicators capable of relaying two-way digital messages up to the Starsys satellites will radiate 2 to 5 watts through a long slender antenna. Cost estimates for the various communicators range from $100 to $350 depending on user-oriented features and design sophistication.

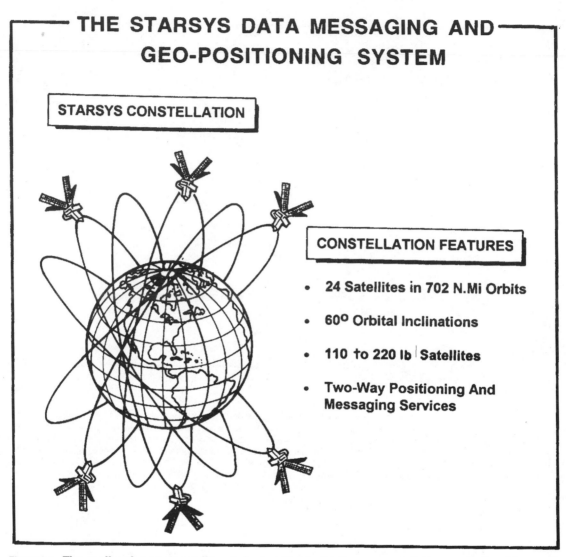

# THE STARSYS DATA MESSAGING AND GEO-POSITIONING SYSTEM

**STARSYS CONSTELLATION**

**CONSTELLATION FEATURES**

- 24 Satellites in 702 N.Mi Orbits
- 60° Orbital Inclinations
- 110 to 220 lb Satellites
- Two-Way Positioning And Messaging Services

**Figure 8.4** The small and compact satellites to be installed in the Starsys constellation are designed to provide nearly continuous geopositioning and message-relay services for users all around the globe. The Starsys constellation, which will consist of 24 satellites in 702-nautical mi orbits with 60 degree inclinations, is being masterminded by Starsys Positioning, Inc. of Lanham, Maryland.

Uplink frequencies are to be in the 150 MHz range with downlinks at about 400 MHz. CDMA modulation techniques will help promote frequency sharing with other digital communication systems. The CDMA encryption techniques will also provide a high degree of privacy for Starsys users. Geographic navigation solutions accurate to about 3300 ft will be provided via a combination of Doppler shift measurements and radio ranging techniques.

The Starsys satellites are being designed to ride into space either as dedicated payloads on a small launch vehicle or as piggyback payloads aboard a larger multistage booster rocket. The cost to build, launch, and begin operating the overall system with an intermediate cluster of five or six satellites is estimated to be on the order of $50 million. Starsys Global Positioning, Inc. is a joint venture between two respected U.S. companies:

1. ST Systems Corporation of Lanham, Maryland
2. North American Collection and Location by Satellite (NACLS, Inc.) of Landover, Maryland

Both commercial entities are located in the greater Washington, D.C. metropolitan area.

## International Constellations

Digital messaging satellites using dirt-simple store-and-forward design techniques have been operated by the military establishments in the United States and the former Soviet Union for many, many years. The Soviets have been especially ambitious in this low-cost messaging arena. Indeed, at their peak, the Soviet Union's command and control functions depended upon at least 40 operational store-and-forward satellites arranged in three low- and moderate-altitude constellations, which are sketched in Fig. 8.5.

### The Former Soviet Union's Store-and-Forward Communication Satellites

Two of the former Soviet Union's store-and-forward satellite constellations employ 74 degree orbital inclinations although, as Fig. 8.5 indicates, their constellation architectures are considerably different.

The three-plane constellation sketched at the top of the figure is made up of three satellites flying around the earth at average altitudes of 432 nautical mi. Each of these three satellites weighs approximately 1650 to 2200 lb.

The second Soviet constellation is characterized by the unusual deployment technique used in scattering the various satellites across the sky. On each launch the Soviets deposit eight separate satellites clustered around the 810-nautical mi altitude regime. Each store-and-forward satellite, which weighs about 110 lb, is placed in an orbit with a slightly different average altitude. The gravity induced by the earth's equatorial bulge then causes the orbit planes of the various satellites to gradually drift apart as indicated by the second sketch in Fig. 8.5.

The third constellation launched by the former Soviet Union consists of two planes of satellites in 82.6 degree orbital inclinations 756 nautical mi high. As Fig. 8.5 shows, the two planes intersect one another at a 90 degree angle. Each of the satellites in these orbits weighs 500 to 550 lb. Simple store-and-forward messaging techniques are used by the satellites which employ passive gravity gradient stabilization to achieve earth-seeking attitude control.

In 1992, a Russian space consortium called SMOLSAT began to market commercial copies of their country's military store-and-forward satellites under the

# THE FORMER SOVIET UNION'S TELEGRAPH SATELLITES

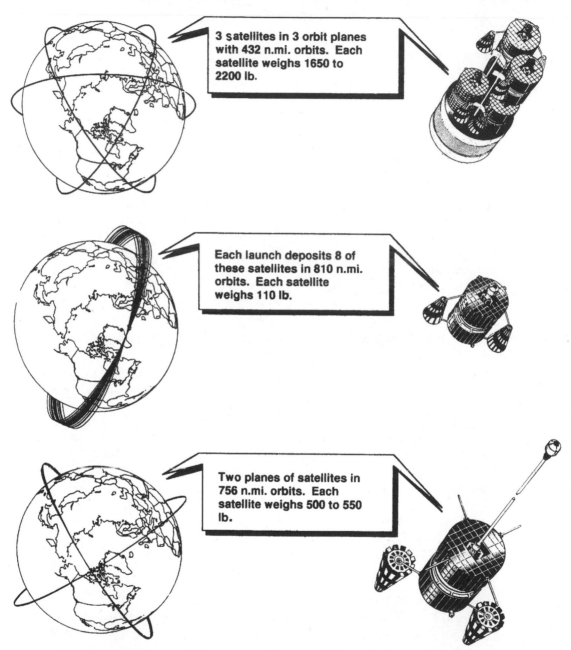

3 satellites in 3 orbit planes with 432 n.mi. orbits. Each satellite weighs 1650 to 2200 lb.

Each launch deposits 8 of these satellites in 810 n.mi. orbits. Each satellite weighs 110 lb.

Two planes of satellites in 756 n.mi. orbits. Each satellite weighs 500 to 550 lb.

**Figure 8.5**    Aerospace experts in the former Soviet Union are currently operating more than 40 store-and-forward telegraph-style communication satellites apportioned among three orbital constellations. Two of the constellations are arranged in conventional geometrical patterns with two and three orbital rings, respectively. The satellites in the third constellation are all launched into the same orbital plane with slightly different average altitudes. Over a period of several weeks, the gravitational perturbations induced by the earth's equatorial bulge begin to create a gradual divergence in the nodal crossing points of their orbit planes.

name "GONETS." A full-fledged system of 30 GONETS satellites is scheduled to come online in 1995. Hand-held communicators weighing less than 5 lb each will access the satellites in the 300 MHz frequency range.

## Leostar's Constellation Design

The Leostar development is being financed by Italspazio in Rome, Italy. The satellites their engineers have begun to design will provide two-way store-and-forward digital messaging and position reporting to users stationed around the globe. The 24 Leostar satellites will be delivered into 432-nautical mi polar orbits where they are expected to provide at least 5 years of useful service.

As Fig. 8.6 indicates, the Leostar satellites will employ gravity gradient stabilization to achieve earth-seeking attitude control. Fine-tuning is accomplished by three mutually orthogonal magnetic torquers which are selectively activated by a dedicated on-board computer. Pointing accuracies of ±1 degree are anticipated by project engineers.

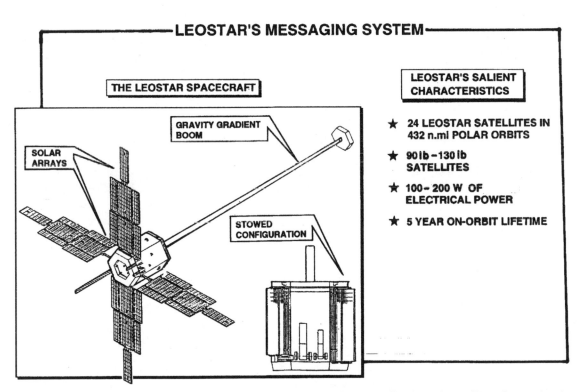

**Figure 8.6**  Leostar is to consist of a spaceborne digital message-relay constellation with satellites that maintain their earth-seeking attitude using gravity gradient stabilization techniques. When the constellation is complete, it will contain 24 Leostar satellites orbiting in 432 nautical-mi orbits inclined 90 degrees with respect to the equator. Each Leostar satellite is rigged with an 8.5-ft gravity gradient boom and four solar arrays arranged in a unique cross-shaped configuration. During launch in its stowed configuration, each Leostar satellite is shaped like a snare drum 30 in. in diameter and about 35 in. tall.

Leostar, which will weigh 90 to 130 lb, will generate somewhere between 100 and 200 watts of electrical power. The satellites are being designed so they can be launched into the desired orbits aboard Scout or Pegasus boosters one at a time or packaged together in pairs. The compact little satellites can also be lofted into space by the European Ariane 4 or Delta boosters in clusters or in the cheaper, but more time-uncertain, piggyback delivery mode.

The gravity gradient boom carried aboard each Leostar satellite is composed of a coiled strip of beryllium and copper. Once the satellite reaches its orbital destination, the boom is automatically deployed vertically upward. The satellite's 3.3-ft messaging antennas are then deployed using spring-actuated hinging mechanisms. Once the antenna swings through an angle of 180 degrees, a latching device rigidly restrains it in its final position.

Four fixed solar panels populated with silicon solar cells are mounted on the two opposite sides of the spacecraft to provide the required electrical power. The on-orbit propulsive burns are powered by liquid propellants consisting of monomethylhydrazine (as fuel) and nitrogen tetroxide (as oxidizer).

## Other Proposed International Constellations

A number of other low-altitude data-only constellations are being proposed by various sovereign entities around the world. The French space agency, CNES, for instance, is studying a global system called TAOS (Greek for "peacock"). If their proposed constellation is launched into space, it will consist of five 330-lb satellites launched into orbits 652 nautical mi above the earth.

The SAFIR constellation with six data-collection and message-relay satellites has been proposed by OHB-Systems at the Advanced Technology University in Brennan, Germany. SAFIR will consist of six satellites orbiting at nominal altitudes of 432 to 640 nautical mi somewhere between 62 and 98 degrees inclination. Doppler shift measurements and Navstar GPS chipsets mounted in the ground communicators would provide highly accurate positioning solutions.

## Simulating the Coverage Characteristics for Various Low-Altitude Constellations

An orbiting satellite is able to cover only a finite surface area (footprint) on the earth. Its instantaneous coverage characteristics can be likened to a flashlight beam shining on a round ball. As the satellite travels around its orbit, the footprint moves along with it to cover different surface regions.

Simulating the coverage for a single low-altitude satellite is extremely easy to do. But, when the satellites in a large constellation are simultaneously covering different portions of the globe, the simulation algorithms can become extraordinarily complicated. Moreover, if a constellation designer includes extra, unnecessary satellites in the proposed constellation, the net waste of resources can quickly accumulate since building and launching any extra satellites will likely cost the constellation sponsors several million dollars each. The analyti-

## FINDING CHEAPER WAYS TO GO ABOUT
## THE BUSINESS OF SPACE

Centers of excellence exist in almost every field. Manhattan has long been a center of excellence for book publishing. And MIT and Cal Tech have, for many years, been centers of excellence for the mechanical engineering of everything from industrial robots to unmanned deep-space probes.

Ever since the earliest days of the American space program, the Applied Physics Laboratory at Johns Hopkins University has been a center of excellence for the design and operation of low-cost satellites. Researchers at that relatively small facility have masterminded such low-cost innovations as gravity gradient stabilization and magnetic momentum dumping. They have also pioneered the broadranging use of precise Doppler shift measurements to achieve instantaneous position solutions. Spacecraft packaging technology, skewed momentum wheels, and a whole host of other simple but innovative techniques for doing more with less all came from that small laboratory nestled among the sugar maples on the outskirts of Laurel, Maryland.

Work there continues apace. So far, the clever engineers on the staff at the Applied Physics Laboratory have not yet directed their full attention toward the development of low-cost space transportation systems. But, when they get around to that challenging assignment, the cost of getting to and from outer space will probably drop by an order of magnitude.

cal methods used in systematically studying and evaluating a constellation's coverage capabilities date back to the 1960s, and beyond. No simple closed-form methods are available for precisely sizing a constellation. Practical simulations for finding the most suitable constellation architecture for a particular application almost invariably involve complicated iterative processes combined with generous portions of human judgment and expertise.

### A Brief Survey of Various Existing Coverage Analysis Techniques

One popular method for studying the coverage capabilities of low-altitude satellite constellations is the so-called "grid search" technique in which a world map is divided into a checkerboard pattern. During the grid search the computer freezes the locations of the satellites in the constellation, then in effect crawls along the surface of the earth from grid point to grid point. When it arrives at each new grid point, it scans the sky to determine how many satellites are within view from that particular location. It then prints that number at the appropriate grid location before moving on to the next grid point in the sequence. Grid searches provide reasonably accurate results, but for fine-meshed grids, they can require an enormous number of time-consuming computer processing steps.

An alternate technique, the "enclosing polyhedron approach," does not yield the point-by-point coverage details provided by a properly constituted grid search, but the polyhedron approach can be used to verify whether or not the entire earth is continuously covered by a particular constellation. This is accomplished by noting that, if the solid polyhedron constructed by connecting the centers of the satellites with a series of straight lines completely encloses the earth, then at that particular instant the satellites will have access to its entire surface.

A third approach for computing the instantaneous coverage characteristics of a constellation is based on the "flashlight" model previously described. Cone-shaped coverage regimes are projected onto the globe one at a time with careful attention directed toward any overlaps on the earth's surface. This approach has been adopted by a number of highly automated routines, including the "Space Eggs" computer simulation program that is described in detail in the next section.

## The "Space Eggs" Computer Simulations

Space Eggs is a computer-based coverage evaluation program that uses the special pixel manipulation capabilities and the color graphics routines built into today's microcomputers to determine the coverage characteristics of large low-altitude satellite constellations. Because it is so fast and efficient and because its color displays are so easy to understand, the Space Eggs routine is ideally suited for use in connection with today's large and complicated mobile satellite constellations.

The Space Eggs simulation is named after the unique geophysical model (earth map) on which the simulation results are displayed. As Fig. 8.7 indicates, the Space Egg is an oval-shaped map of the world called a Mollweide equal-area projection. Notice that the Mollweide projection models the entire surface of the earth, not just the "front" side of it. The Mollweide projection preserves areas. This means that a small circle of fixed size sketched on any part of its flat surface represents the same amount of area on the surface of the spherical earth.

With the fine-grained resolution provided by the video displays that come with today's microcomputers, a Mollweide map projection typically spans about 100,000 pixels. Thus, any satellite footprints laid down on such a map projection are represented with excellent fidelity and resolution.

When the geometrical characteristics of a satellite constellation are fed into the computer, it temporarily freezes the positions of the satellites, then lays their footprints down onto the Mollweide equal-area projection one at a time so the footprints can be stored in the computer. Once all the satellite footprints have been electronically stored in this manner, the program goes back and recalls all the stored projections, one by one.

Each of the 100,000 pixels is initially labeled with a binary zero (white) as a starting value. Then, as the various footprints are laid down on the screen, the binary values (colors) of the affected pixels are increased to higher and higher values wherever a new footprint falls onto their geographical regime. Each binary number represents a different color on the video screen and the colors can all be displayed simultaneously on the screen.

When all of the footprints have been laid down in this manner, the percentage of the overall Space Egg covered by a particular color is identical to the percentage coverage provided by that many satellites. For example, a pixel color

## SIMULATION TECHNIQUES USED BY THE SPACE EGGS PROGRAM

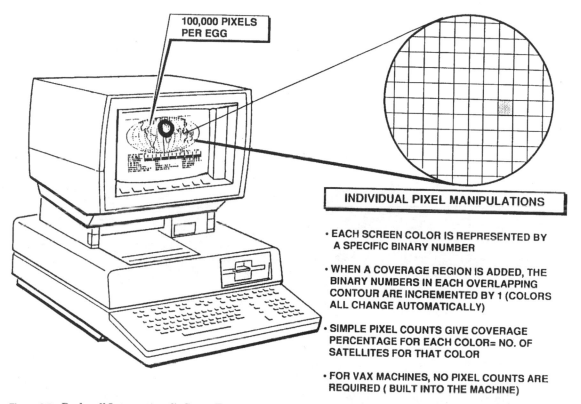

**100,000 PIXELS PER EGG**

**INDIVIDUAL PIXEL MANIPULATIONS**

- EACH SCREEN COLOR IS REPRESENTED BY A SPECIFIC BINARY NUMBER

- WHEN A COVERAGE REGION IS ADDED, THE BINARY NUMBERS IN EACH OVERLAPPING CONTOUR ARE INCREMENTED BY 1 (COLORS ALL CHANGE AUTOMATICALLY)

- SIMPLE PIXEL COUNTS GIVE COVERAGE PERCENTAGE FOR EACH COLOR= NO. OF SATELLITES FOR THAT COLOR

- FOR VAX MACHINES, NO PIXEL COUNTS ARE REQUIRED ( BUILT INTO THE MACHINE)

**Figure 8.7**   Rockwell International's Space Eggs computer simulation program determines the coverage patterns for various low-altitude satellite constellations using the pixel manipulation capabilities built into today's high-speed electronic digital computers. The computer program initiates its simulations by superimposing the coverage area of each satellite onto a Mollweide equal-area projection representing the entire surface of the earth. Color-coding techniques are used to represent the overlap between the coverage contours of the various satellites. Binary pixel counts are then employed to determine the precise coverage characteristics of each proposed constellation.

of one (red) equals coverage by one satellite, whereas a pixel color of three (green) equals coverage by three satellites. As Fig. 8.8 indicates, the circular coverage regions of the various satellites can usually be discerned on the screen—except when extremely large numbers of satellites are being flown in a gigantic constellation.

When all the stored projections have been laid down on the screen, the computer checks the various pixel colors, line by line, while storing (and displaying) the statistical data associated with each individual color. For many engineering studies, the statistics on the instantaneous graphics are of primary concern. But the Space Eggs program can also handle more generalized studies in which the satellite positions are incrementally propagated around their orbits a few degrees at a time. In this mode the program computes the conglomerate coverage statistics for the constellation as a whole as the satellites move through space.

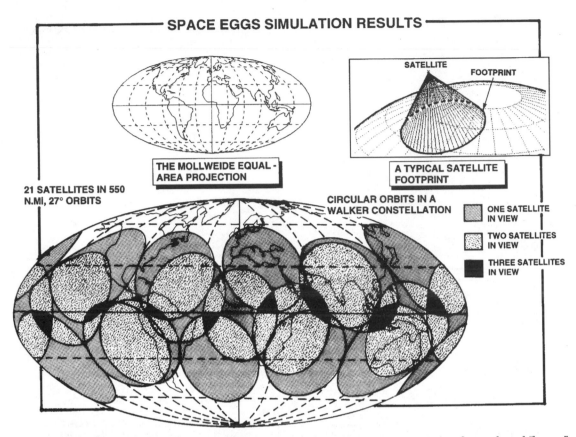

**Figure 8.8**  The simplest coverage patterns for the Space Eggs simulation program consist of cone-shaped "beams" radiating down from each satellite to intersect a Mollweide equal-area projection of the spherical earth. Overlapping areas with various levels of redundant coverage automatically take on the proper color coding. Simple binary pixel counts are then employed to determine what percent of the map is within the overlapping coverage footprints of various numbers of satellites in the constellation. In this simple case, 21 satellites are arranged in 550-nautical mi Walker orbits tipped at an angle of 27 degrees with respect to the equator. Notice that the constellation provides nearly continuous coverage for a latitude band stretching between ±45 degrees north and south latitude.

The Space Eggs computer simulation program was conceived and developed by Janis Indrikis and Bob Cleave at Rockwell International in Downey, California. They have rigged it so it can simulate irregular satellite coverage patterns (donut-shaped regions, wedge-shaped regions). The coverage patterns associated with gravitationally perturbed trajectories and elliptical orbits can also be simulated by the various Space Eggs computational routines.

# Low-Altitude
# Voice Messaging Systems

*It must be considered illusory to think that radio
will ever supersede wire telegraphy.*

FERNINAND BRAUN
*College professor, 1900*

"Standing on a rocky strip of land, somewhere in the middle of the Pacific, you pull out a chunky little telephone from an inside pocket, extend an aerial, and dial a number...." As we amble through that colorful opening sentence, we suddenly begin to realize that the professional wordsmiths on the staff of *The Economist* have found an excellent way to personalize their story of tomorrow's mobile communication systems. "Soon the radio signals bearing your call are dancing directly from your telephone to a necklace of satellites 500 miles or so above the earth, and thence back down to where your great-aunt Madge sits on the other side of the world dispensing sensible advice."

Tomorrow's brave new world of instantaneous communications is careening toward us at a bone-jarring pace now that mobile communication satellites—long the dream of aerospace entrepreneurs—are being converted into genuine hardware waiting for genuine rocket rides into space. Within 2 years or so, clever little *Star Trek*-style communicators will be radiating soft cries for help toward everyone's distant Aunt Madge. And, when they are being sold in every Radio Shack, the diameter of our world will inevitably shrink by another notch or two. "Desert islands will never be the same again," *The Economist* notes, "nor will political dissidents, remote third-world villagers, or peripatetic businessmen."

The sections to follow summarize the technological developments that are pulling us toward the world of tomorrow where all of us will be able to buy, use, and enjoy personal communicators linking us with anyone, everywhere who takes the trouble to become similarly equipped. Our story opens with the Iridium constellation, an ambitious and highly refined swarm of talkative little satellites, being masterminded by Motorola's courageous engineers.

## The 66-Satellite Constellation from Iridium

In June 1990, experts at Motorola, Inc. unveiled their blue-sky vision for a futuristic new constellation of mobile communication satellites. Their initial architectural approach called for seven slender rings of satellites marching in single file up over the North and South Poles with 11 satellites in each circular ring.

Seen from a distant vantage point, planet earth would thus resemble a ball-shaped globule of sugar being attacked by an insistent swarm of honeybees. Each energetic bee would be a communication satellite interlinking busy business travelers equipped with pocket telephones allowing them to reach anyone who lives or works on our beautiful blue planet—or is whirling around it at an orbital altitude of 200 nautical mi or less.

"In essence, Iridium is a cellular telephone system turned upside down," notes James R. Asker from *Aviation Week and Space Technology*. "Instead of users moving through cells established by ground stations, the cells move overhead as the satellites pass over the users."

Motorola engineers decided to name their mobile communication system "Iridium" because their 77-satellite constellation was in direct analogy with the 77 electrons circling around iridium's nucleus. Iridium is one of the platinum metals, a precious silver-white substance harder than iron, nearly as brittle as flint glass, and more dense than copper or brass. Iridium alloys are sold as jewelry. They are also used in surgical pins and pivots, as fountain pen tips, for electrical contacts and sparking points, and as bearings in high-precision navigational compasses.

Iridium has also been discovered as a thin, unexpected layer laid down in worldwide sediments 65 million years ago. Its presence helped spark a controversial theory that land-based dinosaurs became extinct—with geological swiftness—when a large meteorite slammed down onto the earth, thus creating a sun-choking cloud of dust in the upper atmosphere.

To honor the colorful name they chose, Motorola's marketeers send out public relations brochures in a polished silver folder with big letters spelling "IRIDIUM" embossed on the front cover. Unfortunately, when they later decided to increase the number of spot-beams and the transmitter power of each satellite, they also decided to reduce the number of satellites to only 66.

The element with 66 electrons whirling around its nucleus is dysprosium, so no one can blame Motorola's marketeers for not renaming their satellite-based system "Dysprosium," a name that has none of the significance, pronounceability, or sex appeal of "Iridium."

Dysprosium is a rare-earth metal whose Greek name means "hard to get." It is often found in association with erbium and holmium and other similarly obscure rare-earth substances. When cooled to a sufficiently low temperature, dysprosium is strongly attracted by a magnet. When used as the name for a new constellation of satellites, it leaves everyone, including potential financiers, feeling flat and unresponsive.

Iridium is the biggest, most complicated venture in Motorola's 66-year history. So far, Motorola has ponied up most of the initial cash required, but company officials are hoping to reduce their ownership fraction to only 15 percent as more and more well-heeled participants begin to come online. Present plans call for a positive cash flow by 2002, 4 years after startup of the full Iridium constellation.

### Iridium's Constellation Architecture

The Iridium constellation, with a company-projected price tag of $3.4 billion, is one of the most costly concepts ever devised for providing mobile communication services to a globally distributed class of users. The Iridium satellites, which are to travel around the earth in 413-nautical mi polar orbits, will weigh about 1600 lb each. As the sketches in Fig. 9.1 indicate, the 66 satellites will be equally spaced within their six orbital rings.

Each satellite in the Iridium constellation will send out 48 pencil-thin spot-beams each of which can handle 230 simultaneous duplex conversations. Irid-

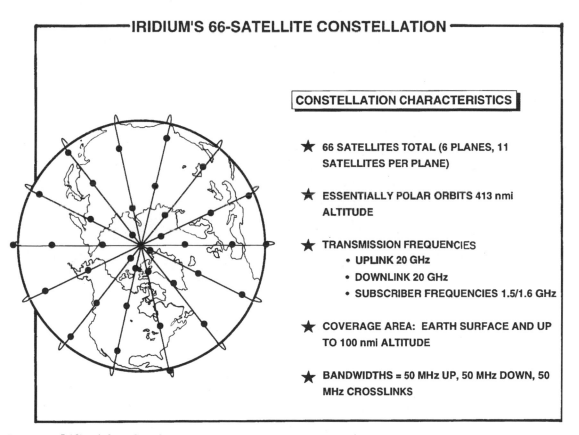

**IRIDIUM'S 66-SATELLITE CONSTELLATION**

**CONSTELLATION CHARACTERISTICS**

★ 66 SATELLITES TOTAL (6 PLANES, 11 SATELLITES PER PLANE)

★ ESSENTIALLY POLAR ORBITS 413 nmi ALTITUDE

★ TRANSMISSION FREQUENCIES
  • UPLINK 20 GHz
  • DOWNLINK 20 GHz
  • SUBSCRIBER FREQUENCIES 1.5/1.6 GHz

★ COVERAGE AREA: EARTH SURFACE AND UP TO 100 nmi ALTITUDE

★ BANDWIDTHS = 50 MHz UP, 50 MHz DOWN, 50 MHz CROSSLINKS

**Figure 9.1** Iridium's low-altitude voice messaging constellation will consist of 66 polar orbiting satellites symmetrically arranged in 11 orbital rings with 6 satellites in each ring. The satellites, which will trace out circular 413-nautical mi orbits, will each send down 48 spot-beams with an average diameter of 363 nautical mi. Working in concert, the 66 Iridium satellites will cover the entire surface of the globe plus a thin spherical shell surrounding it at an attitude of about 100 nautical mi.

ium's highly capable satellites will feature on-board switching and crosslink message relay from satellite to satellite. Well-designed intersatellite links will give the Iridium constellation unmatched capabilities to locate the source of each call and complete the necessary connection within 10 seconds. This is to be accomplished electronically by passing the call directly from satellite to satellite until it reaches the desired destination.

The overall architecture for the Iridium constellation is summarized in Fig. 9.2. Notice how the schematic diagram highlights the Ka-band satellite-to-satellite crosslinks at 20 GHz and the Ka-band downlinks connecting the Iridium satellites with their gateway stations and their ground-based system control stations. The transmission links connecting the hand-held communicators, the paging units, and the remote area telephones will all be handled with the same L-band frequencies between 1.5 and 1.6 GHz.

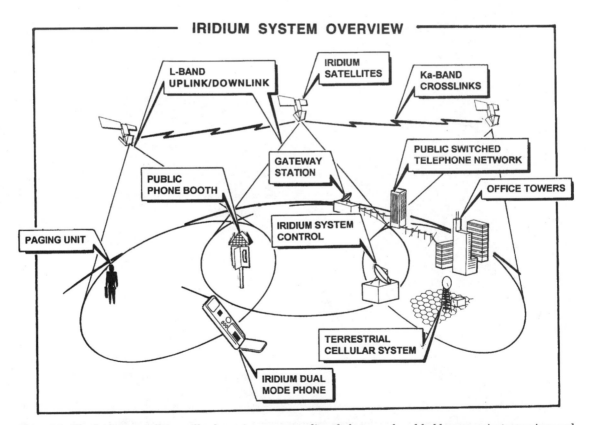

**Figure 9.2**  The Iridium satellites will relay voice messages directly between hand-held communicators using modulated L-band signals in the 1.5 to 1.6 GHz range. The satellite-to-satellite crosslinks are operated in the Ka portion of the frequency spectrum at 20 GHz. Similar Ka-band transmissions are also used to connect the Iridium satellites with their gateway stations and their ground-based system-control stations. When a call is initiated, the dual-mode communicators will automatically check to see if a terrestrial cellular telephone connection is available. If not, a more expensive satellite relay link will be set up to handle the desired voice messaging connection.

## Signal Modulation Techniques and Frequency Selections

Unlike most of its spaceborne competitors which rely on spectrum-sharing using CDMA modulations, Iridium employs TDMA architecture. This approach will require that a dedicated portion of the frequency spectrum be allocated to Iridium to provide interference-free operation.

The necessary frequency allocations will be difficult to obtain from domestic and international regulatory bodies, but if the spectrum is obtained Iridium offers one overriding advantage to ground-based users. "Iridium will provide subscribers with units that offer high-quality sound anywhere, including indoors," says Motorola's vice chairperson John F. Mitchell. "The others will have to be used *outside.*"

On-board switching and crosslink messaging techniques add quite a bit of complexity to the Iridium satellites, thus increasing the probability of on-orbit failures. However, both on-board switching and crosslinking have been demonstrated on a number of military and civilian communication satellites, including the Department of Defense's Milstar, NASA's Advanced Technology Satellite, and the Navstar GPS.

Iridium's transmission rates have been set at 4800 bits/second for voice, and both 4800 and 2400 bits/second for digital data transmissions. Computer processing techniques using vocoders will be employed to compress each conversation onto the allocated bandwidth of 4800 bits/second. Fifteen to 20 widely distributed gateway stations will be installed all over the world to provide the necessary interfaces with the public switched telephone networks.

In 1992, the FCC awarded Motorola an experimental license to construct and launch five Iridium satellites to demonstrate the feasibility of their overall system concept. These launches are scheduled to begin in 1996.

## Spacecraft Design Techniques

Each 1600-lb Iridium satellite is rigged with two winglike solar arrays protruding from the sides of the spacecraft. The earth-seeking main body of the spacecraft carries the L-band communication antennas and the Ka-band antennas, which are used to communicate with the various other satellites in the constellation and the gateway stations on the ground.

## Iridium's Communicators

When an Iridium communicator is activated, the nearest satellite (working in concert with the ground-based Iridium network) ascertains the validity of that subscriber's account, then determines the location of the user. The system automatically checks to see if an inexpensive terrestrial link is available to handle the call. If not, the call is relayed through the nearest satellite and, if necessary, from satellite to satellite to its destination.

If an Iridium subscriber is at a remote location, the call will be transmitted directly to the intended recipient. If the subscriber is in the vicinity of a land-based telecommunication system, conventional terrestrial communication channels will be used instead. Navstar GPS chips built into the hand-held communicators help position the users for switching and billing purposes.

The communicators can be rigged to handle any type of telephone transmission—voice, data, paging, or fax. The initial batch of communicators will be available in three fundamentally different models:

1. Paging units capable of receiving digital pulse trains only—no voice

2. Single-mode voice messaging units capable of sending and receiving calls relayed only through the Iridium satellites

3. Dual-mode units that will automatically select a terrestrial cellular relay—when it is available—or switch to a satellite relay whenever terrestrial links are not available

A prototype dual-mode communicator is depicted in Fig. 9.3. Notice that it provides both voice messaging and liquid crystal alphanumeric displays. This unit is roughly the same size and weight as one of today's analog cellular telephones, but it is all-digital so it will not eat batteries the way today's hand-held cellular telephones do. The antenna for the unit is also shown in Fig. 9.3. Notice that it is only slightly larger than the ballpoint pen at the bottom of the picture.

### International Financial Arrangements

The Great Wall Industry Corporation is scheduled to launch 20 of the 1600-lb Iridium satellites using upgraded Chinese Long March 2C boosters. Krunichev Enterprises of Moscow will share Iridium's launch orders with China's Great Wall and McDonnell Douglas of St. Louis, Missouri, which will use its Delta 2 boosters to orbit the Iridium space vehicles. Each Russian Proton booster will carry seven Iridium satellites; each McDonnell Douglas Delta 2 booster will carry five; and each Chinese Long March booster will carry two.

The satellites are being designed in part by Motorola's Satellite Communication Division in Scottsdale, Arizona and are being constructed by Lockheed at Calabassas, California. The early communicators are being built by Japan's Matsushita Electric (Panasonic) under a small-scale contract.

Company officials are hoping to sign up 2 million Iridium customers by the year 2002. Of those 2 million users, 700,000 are expected to be pure paging users who will receive short bursts of digital data only.

Through aggressive marketing campaigns Iridium officials have obtained pledges for more than $800 million in backing for their mobile communication system. They have also managed to set up barter agreements to ensure that their satellites will be able to hitch enough rides into space. Motorola has come a long way toward international acceptance of an advanced telecommunications concept that, in the early design phases, seemed entirely fanciful to many people. Some of these people are now investing in the project.

Daini Denden and Kyocera Corporation, both in Tokyo, control a 15 percent stake in the corporation and two Saudi Arabian groups led by the Mawarid

# HAND-HELD COMMUNICATORS FOR THE IRIDIUM PERSONAL COMMUNICATION SYSTEM

**Figure 9.3**  Iridium's personal communicators, which are about the same size as today's hand-held cellular telephones, will use digital time division multiple access techniques to send and receive voice messages and digital data to or from any point on the surface of the globe. The attractive communicators are slated for delivery in one of three different commercial models: (1) paging units capable of receiving digital data only, (2) single-mode units that can relay messages through the Iridium satellites, and (3) dual-mode units that automatically select terrestrial cellular telephone links when they are available before defaulting to the more expensive satellite-based communication links. (*Courtesy of Iridium.*)

## MOTOROLA'S INDUSTRY SUCCESS

"Even today half the world's population lives more than two hours away from a telephone," observes *The Economist* in appropriately worrisome tones, "and that is one reason why they find it hard to break out of their poverty."

But if Motorola's entrepreneurial innovators have their way, people everywhere will soon have to decide whether or not they really do want to communicate. Cutting-edge communication technology has long dominated the thoughts and deeds of company leaders at the Motorola Corporation. When he founded the company in 1928, Paul Galvin knew that his first product, the "battery eliminator," might soon become obsolete. So he sketched preliminary plans to make and sell car radios, which appeared to have a much more promising future.

From that tentative beginning, the company Paul Galvin founded segued into two-way radios, World War II's famous walkie-talkies, and then, when the war was over, into television. Soon company marketers were busy selling the first solid-state television sets under the Quasar brand name.

Of course, Motorola today occupies a commanding position in the cellular telephone field. In 1985, an engineering team, armed with the new integrated circuit, redesigned their company's most popular cellular telephone. They ended up with a sleek little unit that "had 70 percent fewer components, was two-thirds smaller and lighter, and could be assembled largely by robots," observed *The Wall Street Journal*. Moreover, the new, improved cellular telephone "took one-tenth as long to assemble as before, and defects were reduced 90 percent."

Today Motorola is the preeminent leader in the production of cellular telephones, and their marketers have been branching out into all forms of wireless communications, including personal pagers. In 1991, Motorola shipped 100,000 pagers to China. In 1992 they shipped 1 million. In 1993 they shipped over 3 million.

Unlike many high-technology companies, where engineering expertise is not an important criterion on the promotional ladder, at Motorola almost every single top executive is a trained engineer. "As fast as technology moves today, there's an advantage if the leadership of the company is comfortable with technology," remarks Motorola chairman George Fisher. "You have to have a basic love of technology."

Fisher, who holds a Ph.D. in applied mathematics, also believes that technical challenges, not pay rates, are what motivate engineers. "People must be challenged to achieve what on a day-to-day basis they might have thought was unachievable," he observes. That simple sentiment might well become the basis for a raising new company song at Motorola, a song that could call on America to become far more competitive with other entrepreneurs throughout the rest of the industrial world.

Group of Riyadh jointly own 15 percent. Several telecommunications companies have signed up for 5 percent. These include Sprint Corporation of Kansas City, Missouri; BCE Mobile, a subsidiary of BCE, Inc. of Montreal, Canada; Societa Finanziaria Telefonica Per Azioni of Rome, Italy; and the United Communications Industry Company of Bangkok, Thailand.

Other investors include China's Great Wall Industry Corporation of Beijing and the Russian Federation's Krunichev Enterprises of Moscow, both of which will be launching some of the Iridium satellites into their low-altitude orbits. The two companies have each bartered for a 5 percent share of the corporation.

"In addition to the money [these pledges] bring to foster system development, the commitments put the reputation of major international industrial organizations behind what, a few years ago, seemed a fantastic dream," observes James T. McKenna of *Aviation Week and Space Technology*. His thoughts are echoed by executive John Mitchell of Motorola, who is proud of the money his company has managed to raise. "This is a milestone allowing us to clearly see the beginning of a new class of global wireless communications before the 21st century," he notes with a big grin.

## The Globalstar Constellation

Globalstar is a potential competitor for Iridium, but it uses a much simpler constellation architecture with gateway stations tied into the existing ground-based infrastructure to be used for message switching.

The Globalstar constellation is being planned with 48 satellites (including 8 on-orbit spares) boosted into eight orbit planes inclined 52 degrees with respect to the equator. The nominal altitude of the satellites is 750 nautical mi. Each of the 490-lb Globalstar satellites will employ six spot-beams with CDMA modulations for highly effective use of the available frequency spectrum.

### Globalstar's Constellation Selection Rationale

Communication engineers at Loral Cellular Systems purposely avoided the use of crosslinking between the Globalstar satellites and on-board switching techniques because they are convinced that these approaches are needless technological frills driven by engineering considerations rather than any real user needs.

"It makes little sense to use expensive satellite capacity, as Motorola does, to bypass a terrestrial network that contains an abundance of low-cost, wide-band, highly reliable, long-haul transmission facilities," argues Dale N. Hatfield of Hatfield Associates in Boulder, Colorado. He reached that unshakeable conclusion when his company was commissioned to compare the overall architectures being employed by the two constellations.

Hatfield's line of argument helps explain why the overall architecture for the two mobile communication systems is grossly different despite the fact that both of them are being designed to serve essentially the same clientele. The use of ground-based switching greatly simplifies the Globalstar satellites' design and construction. As Hatfield concludes, "The Globalstar satellites can be constructed as simple 'bent pipe' repeaters and, consequently, are far less complex than the Iridium spacecraft."

Hatfield also has other bones to pick with Iridium's constellation architecture. The decision to go with polar orbits, he concludes, was driven by the desire to simplify the crosslinks used by the Iridium satellites. Each satellite can crosslink with its adjacent neighbors in the Iridium constellation. Polar orbits help make the crosslink geometry simpler, but crosslinking also to some extent results in bad coverage geometry. "In order to simplify the problem of providing crosslinks," Hatfield writes, "the Iridium satellites will be placed in polar orbits, ensuring that the satellites will wastefully and unnecessarily cover the polar regions."

The Globalstar constellation with its smaller satellites and its simpler constellation, by contrast, relies on inclined orbits "to provide significantly improved coverage efficiency." The Globalstar approach permits company engineers to provide only the skimpiest coverage for the largely unpopulated polar regions while striving to achieve much improved coverage for the more densely populated high-latitude areas on both hemispheres.

Sketches and additional background information on the Globalstar constellation are presented in Fig. 9.4, which also includes a three-dimensional "snapshot" showing where the satellites will be positioned with respect to one another at a particular instant in time. Communicators for the system will probably retail for about $750. Each minute of "talk time" is slated to cost about 30 cents.

## Spacecraft Design Features

The 490-lb Globalstar satellite features a flat, elongated center-body structure flanked by two flat-panel silicon solar arrays that provide 150 watts of electri-

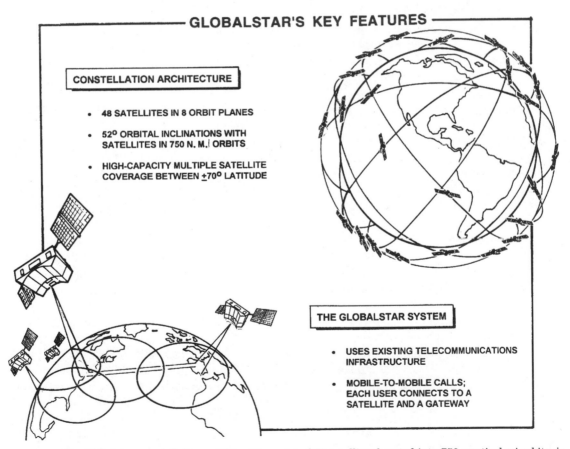

**GLOBALSTAR'S KEY FEATURES**

**CONSTELLATION ARCHITECTURE**

- 48 SATELLITES IN 8 ORBIT PLANES
- 52° ORBITAL INCLINATIONS WITH SATELLITES IN 750 N. M. ORBITS
- HIGH-CAPACITY MULTIPLE SATELLITE COVERAGE BETWEEN ±70° LATITUDE

**THE GLOBALSTAR SYSTEM**

- USES EXISTING TELECOMMUNICATIONS INFRASTRUCTURE
- MOBILE-TO-MOBILE CALLS; EACH USER CONNECTS TO A SATELLITE AND A GATEWAY

**Figure 9.4**  The Globalstar constellation, which will consist of 48 satellites boosted into 750-nautical mi orbits, is structured to provide ample coverage for a band stretching above and below the equator between ±70 degrees latitude. Globalstar's three-axis stabilized satellites, each of which weighs 490 lb, feature two flat-panel solar arrays and a catamaran-shaped central body where the satellite's bent-pipe transceivers are located.

cal power averaged over the on-orbit life of each satellite. The ultimate constellation will include 48 satellites capable of providing nearly continuous global coverage.

However, constellation engineers may install an interim 24-satellite constellation whose prime service area would be the continental United States. The Globalstar constellation, when fully implemented, will provide 28,000 simultaneous voice and data channels each with a 4800 bits/second data rate. The antenna patterns will be formed on board each satellite by flat arrays constructed from both active and passive antenna elements.

## Globalstar's Sponsors

The Globalstar constellation is being financed by Loral Qualcomm Satellite Services, Inc., a company that was founded specifically to handle the Globalstar voice messaging system. The Loral Qualcomm company was founded by combining key elements from the Loral Corporation and Qualcomm, Inc.

Loral is a leader in defense and spaceborne communication systems with headquarters in New York. Qualcomm, Inc. is a leading-edge technology and communications company located in San Diego, California. Together the two commercial entities are hoping to complete their proposed low-altitude voice messaging system for a total cost of about $1.7 billion.

The constellation they are building is designed to concentrate its prime coverage beams between ±70 degrees latitude. From its 750-nautical mi altitude each Globalstar satellite can provide coverage for a region 2600 nautical mi in diameter. Approximately 125 ground stations will link the Globalstar satellites into the conventional telephone networks spotted around the globe.

Globalstar's communicators will be similar in size and cost to the digital cellular telephones now being introduced in a few heavy-traffic areas such as Manhattan and the suburbs of Los Angeles, California. Transmission power will be only about 0.2 watt—or roughly 1 percent of the power level of an outdoor Christmas tree bulb. Most of the personal communicators capable of accessing the Globalstar satellites will be rigged as dual-mode devices capable of accessing terrestrial cellular telephone systems too.

The Globalstar gateway stations will provide radionavigation services to fix the positions of self-selected users. These measurements, which are made by monitoring two successive round-trip signal travel times, will provide an accuracy of about 1000 ft.

Soft hand-offs using path diversity will enhance the performance of the Globalstar system. The soft hand-off approach provides robust communication links and enhanced signal-to-noise ratios for successful message transmission even in noisy environments with excessive multipath reflections.

## The Aries Constellation Approach

The 48-satellite Aries constellation being sponsored by Constellation Communications, Inc. is similar in concept to the one being planned for Globalstar. The

275-lb Aries satellites are to be placed in 550-nautical mi polar orbits in four separate orbital plans. Each satellite will draw 107 watts averaged over an on-orbit lifetime of 5 years. Uplinks are to be in the 1.6 GHz (L-band) portion of the frequency spectrum with downlink frequencies at S-band near 2.5 GHz.

The complete Aries constellation will probably cost $400 to $500 million. Talk time is expected to cost about 30 cents for each minute the two-way circuit is used. Aries will provide digital data and voice messaging services at 4800 bits/second together with navigation positioning accurate to 4 or 5 nautical mi. The radionavigation solutions use digitally modulated two-way signals relayed by the Aries satellites. Dramatically improved positioning accuracies can be achieved by inserting special Navstar GPS positioning chips into the Aries communicators.

## Russia's Signal Satellite System

Figure 9.5 summarizes some of the key features of Russia's proposed 48-satellite voice messaging system called "Signal." The satellites in the Signal con-

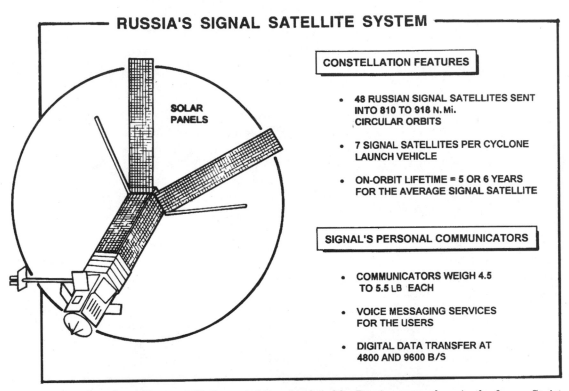

**RUSSIA'S SIGNAL SATELLITE SYSTEM**

SOLAR PANELS

**CONSTELLATION FEATURES**

- 48 RUSSIAN SIGNAL SATELLITES SENT INTO 810 TO 918 N. Mi. CIRCULAR ORBITS
- 7 SIGNAL SATELLITES PER CYCLONE LAUNCH VEHICLE
- ON-ORBIT LIFETIME = 5 OR 6 YEARS FOR THE AVERAGE SIGNAL SATELLITE

**SIGNAL'S PERSONAL COMMUNICATORS**

- COMMUNICATORS WEIGH 4.5 TO 5.5 LB EACH
- VOICE MESSAGING SERVICES FOR THE USERS
- DIGITAL DATA TRANSFER AT 4800 AND 9600 B/S

**Figure 9.5**  The Signal satellite constellation now being developed by Russian researchers in the former Soviet Union is being planned for 48 satellites scattered around circular orbits 810 to 918 nautical mi high. Cyclone boosters are slated to carry seven Signal satellites on each flight toward their intended orbital destinations. Personal communicators for the Signal system will be rather bulky and heavy. Initial versions are being designed to weigh 4.5 to 5.5 lb each.

stellation are to be placed in circular orbits ranging in altitude from 810 to 918 nautical mi. The Signal satellites, which are to be carried aloft in batches of seven aboard the Russian Cyclone booster, will feature four flat, prong-shaped solar arrays cocked at shallow angles and pointed generally upward away from the earth.

The 48 satellites in the Signal constellation will serve as bent-pipe relay stations to provide voice messaging communication links for small ground terminals weighing 4.5 to 5.5 lb each. Launches for the Signal satellites are slated to begin near the end of 1994. The constellation will be considered operational when 24 of the 48 satellites are successfully relaying telephone conversations through space. The other 24 satellites will be launched shortly thereafter, thus providing a more robust constellation with improved dual-satellite coverage capabilities.

In addition to providing voice messaging services for the newly independent Russian states, involving the use of rather large and bulky hand-held communicators, the Signal constellation will support computer links, pagers, and fax machines. Transmissions will be shut down whenever the satellites pass over North America. This will help avoid what Russian designers have called a "legal morass."

## Ambitious Plans from the Teledesic Corporation

The driving force behind the satellite constellation being touted by Edward Tucks and his colleagues at Teledesic is the ongoing shortage of telephone circuits in many underdeveloped nations. According to the International Communications Union: "There are 40 million people on the waiting list for telephone service around the world." Of course, a much larger number of others do not even bother to attach their names to waiting lists.

Company experts at Teledesic (Kirkland, Washington) are totally convinced that they can harness advanced space technology on a grand scale to handle those unmet needs for more and better telecommunication services. Their research indicates that huge numbers of potential subscribers in developing countries cannot get telephone services despite the fact that they have the ability to pay.

"The richest 10 percent of Indonesian households have an average income equal to that of their counterparts in Portugal," notes Edward Tuck, "and they consist of twice as many households as the whole of Portugal....Yet Portugal has twice as many telephone lines as the whole of Indonesia."

Once he scratched out a few pencil-and-paper calculations of that type, Tuck began to figure out how he could fill those unmet needs—on a worldwide scale—with worldwide profits dancing through his head. He has never intended to supply telephone connections directly to the end users. Instead, his company would earn its revenues by supplying global telephone links to the existing telecommunications providers now spotted around the globe. They, in turn, would fill the needs of local customers with messages relayed through the Teledesic satellites.

Edward Tuck had been working on his huge constellation of satellites with a small team of engineers for many years under the name "Calling Communications" when, in March 1994, two successful entrepreneurs, Bill Gates at Microsoft and Craig McCaw, who had founded McCaw Cellular, decided to share his dream. Soon they announced that, over the next 7 years, their new company, called the Teledesic Corporation, would be building and launching huge swarms of flower-shaped satellites. Appropriately, their plan calls for worldwide service to begin in 2001.

"Think of it as a global Internet," says company president Russell Daggett, referring to the highly popular computer-based communications network. In addition to voice and data distribution, the system will also provide video conferencing between any two points on earth.

Gates and McCaw together own 30 percent of Teledesic, whereas Edward Tuck owns 10 percent. As new partners come onboard, their respective shares will be diluted proportionately. Bill Gates cofounded Microsoft, and helped build it into a software behemoth. His personal fortune has been pegged at $7 billion, making him the second richest man in the United States. Craig McCaw built up a vast personal fortune in the cellular telephone industry. His company is being acquired by AT&T for a $12.6 billion share exchange.

## Constellation Architecture

The network of satellites being planned by Teledesic engineers are to be placed in 21 orbit planes each containing 40 active satellites with, in addition, as many as four operational spares in each orbit plane. The 1750-lb satellites will orbit the earth at an altitude of 378 nautical mi in sun-synchronous orbits that will be properly spaced so personal hand-held communicators on the ground will have access to at least two satellites at all times. Dual-access will allow load sharing among satellites to cover for outages and help optimize resource utilization. Current estimates indicate that the system will cost at least $9 billion before the satellites are all in place.

Why does the constellation include such a large number of satellites? Edward Tuck sees his organization "not as a mobile satellite company, but as a *phone* company." His primary intent is to provide reliable voice messaging services to provincial and rural areas of developing countries where conventional phone company services do not have a highly developed infrastructure. He also intends to provide both basic and enhanced services to rural areas in high-income developed countries.

Teledesic will not serve any of the subscribers directly. Instead, the company will "sell the service through the existing government or phone companies." Telephone calls relayed through the satellites will probably cost about 30 cents per minute. With a user base of only 2 million customers, the Teledesic Corporation could just about break even in the first few years. But it soon becomes clear to anyone who talks with them that company marketers have their sights set on much higher goals than that. "Half the world has never made a phone call," notes James Stuart, satellite system manager for the Teledesic Corporation. "Those are the high-growth areas of the world, and this system would be the cheapest way to lay down telecommunications infrastructure in developing countries."

### The Teledesic Satellites

As Fig. 9.6 indicates, the satellites in the 840-satellite Teledesic constellation resemble big, awkward sunflowers with their Ka-band phased-array antennas drooping down toward the ground. The large flat solar array sitting atop each satellite also acts as a cheap parasol to shield the main body of the spacecraft from the burning rays of the sun. The satellites are launched into sun-synchronous orbits at an inclination of 98.2 degrees.

Transmission frequencies were selected in the Ka-band at 20 to 30 GHz. Rain attenuation can sometimes be rather severe in that portion of the frequency spectrum, but the path diversity provided by such a large number of satellites and the high elevation angles (usually 70 degrees or higher) combine to minimize rain attenuation problems. The Ka-band signals will not penetrate foliage or artificial structures so the communicators—or their antennas—will have to be positioned on the outside free of appreciable blockages. A total bandwidth of 200 MHz for the uplinks and 200 MHz for the downlinks will be required for a competitive communication system operated on the profitable scale being envisioned by Teledesic's mission engineers.

## WHAT IS THE LARGEST CONSTELLATION EVER LAUNCHED INTO SPACE?

A *constellation* is a group of similar satellites working together in partnership to provide a useful service. The overriding reason we choose to install a constellation is to get more or better coverage. The Transit navigation satellites are a constellation. Most of the time five or six of them are tracing out polar orbits as they broadcast navigation signals to mariners all around the world. The Inmarsat communication satellites are a constellation. Typically, four of them provide two-way voice messaging and digital data relay for thousands of ships at sea.

What is the largest constellation ever launched into space? My orbital mechanics students enjoy trying to devise clever answers to that deceptively simple question. If we arbitrarily exclude the 1.3 billion whiskerlike copper dipoles that made up Project West Ford, the 840-satellite constellation being planned by the Teledesic engineers is a strong contender for the championship title. Theirs will also be the heaviest constellation ever launched into space. If we assume a constellation of 840 satellites (not counting the 80 some-odd on-orbit spares) each weighing 1750 lb, then the mass of their constellation equals 1.5 million lb.

Actually, the dozen or so Saturn-Apollo payloads launched into 100-nautical mi parking orbits on their way to the moon had a bit bigger total mass. But those payloads do not count because they are not in orbit at the same time. We should give a rousing cheer to the efforts of Edward Tuck and his colleagues at the Teledesic Corporation because they might possibly bring a new space record to America, a record that will not likely be broken by the rocket scientists of any other country for many years.

# TELEDESIC'S FLOWER SHAPED SATELLITES

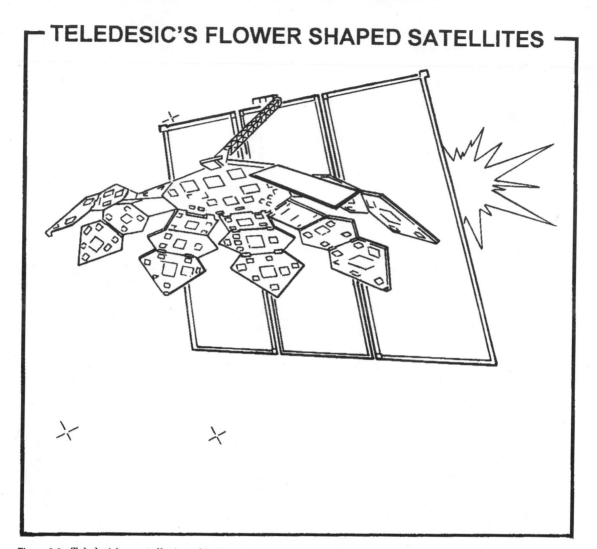

**Figure 9.6**  Teledesic's constellation of 840 active satellites in 378-nautical mi orbits is intended to provide conventional telephone services for large numbers of unserved clients throughout the underdeveloped world. A highly ambitious undertaking conceived on a grand scale, the Teledesic concept will require expenditures of at least $9 billion between now and 2001 when the full constellation of Teledesic satellites is slated to provide operational services to a worldwide clientele.

## Coverage Footprints for the Teledesic Constellation

Typical coverage footprints for the 840-satellite Teledesic constellation are sketched in Fig. 9.7. Advanced aerospace technologies have been employed throughout the system design—active phased-array antennas, fast packet switching techniques, adaptive routing techniques, and *Star Wars*-type lightsat spacecraft components.

The multibeam antennas mounted on each satellite will lay down ground-based coverage patterns that are partitioned into "supercells," each of which measures 86.4 nautical mi on a side. These supercells, which are essentially square, are in turn subdivided into nine cells each. Each cell is 28.8 by 28.8 nautical mi on a side.

As the satellite sweeps across the sky, the cells and supercells are not pulled along with it. Instead, the satellite automatically swivels as it flies to compensate for its forward motion. Consequently, the same spacecraft antennas illuminate the fixed cells on the ground for several minutes at a time.

"Instead of moving with the satellite footprint, the system's cells are arranged in a fixed grid on the earth to which the satellites electronically steer their antennas as they pass," Edward Tuck passionately explains. "This per-

## COVERAGE FOOTPRINTS FOR THE TELEDESIC SATELLITES

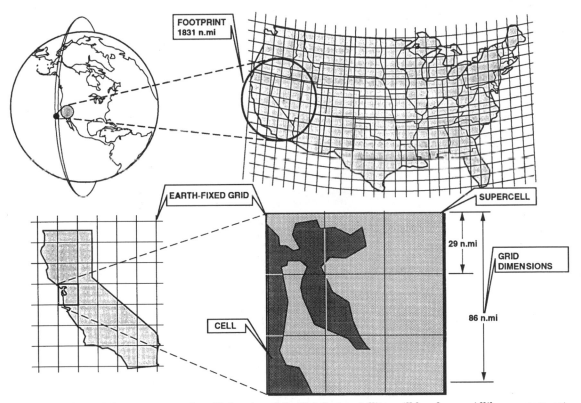

**Figure 9.7**  The multibeam antennas installed on board the Teledesic satellites will lay down gridlike coverage patterns that are divided into essentially square "supercells" measuring 86.4 nautical mi on a side. Each supercell is further subdivided into nine smaller cells measuring 28.8 by 28.8 nautical mi. As a satellite sweeps overhead, it does not maintain a constant earth-seeking orientation. Instead, it gradually swivels to illuminate the same cell regions on the ground below. This dynamic method of satellite attitude control reduces the number of hand-offs required to help make sure the user typically experiences seamless connections for most conversations of average duration.

mits a terminal to keep the same channel assignment for the duration of the cell regardless of the number of satellites involved. Moreover, hand-offs become the exception, rather than the rule."

## Services to Be Provided to the Globally Distributed Users

The 840-satellite constellation being masterminded at the Teledesic Corporation will provide 2 million simultaneous duplex voice channels to serve a 20-million customer subscriber base. A customer base that large is equivalent to half the individuals in the world who have placed their names on waiting lists for conventional telephone services. At a rate of 25 to 50 cents per minute, they will receive toll-quality voice and data services at any location on the surface of the earth.

Half the world's population has never made a phone call. But a lot more of them will be receiving regular phone services in their homes and at their businesses if Edward Tuck, Bill Gates, and Craig McCaw manage to install their fantastic Teledesic constellation.

# 10

# Medium-Altitude Constellations

*They will never try to steal the phonograph—it is*
*not of any commercial value.*

THOMAS ALVA EDISON
*Inventor of the phonograph, 1915*

Communication satellites are almost always headed toward extremely high or extremely low orbital destinations. The ones targeted for high-altitude destinations are most often sent into geostationary orbits at the 19,300-nautical mi altitude—a special orbital location where they can be made to remain motionless with respect to the spinning earth. The ones bound for low-altitude destinations are usually installed in large constellations barely skimming over the ground below.

Sandwiched between these orbital extremes is a vast, featureless domain called "the Abyss" with a number of potentially attractive characteristics. Some experts are convinced that mobile communication satellites sent into the largely unpopulated Abyss can be formed into medium-altitude constellations with some highly effective communication capabilities.

John Draim, senior technologist at Arlington, Virginia's Space Applications Corporation, has spent an appreciable fraction of his professional career analyzing the capabilities of various types of satellite constellations. Draim, in fact, holds a U.S. patent on the only four-satellite constellation known to provide continuous coverage for the entire surface of the globe. His painstaking studies have convinced him that the unpopulated mid-altitude regime holds great promise for tomorrow's constellations of mobile communication satellites.

"It is becoming apparent that the Abyss, the mid-altitude region of space, is virtually unused territory," he notes. "Yet it is also the region in which the potentially most efficient constellations (from the standpoint of coverage versus cost) are to be found."

John Draim is particularly enamored with the enhanced performance capabilities of satellites swinging up through the mid-altitude Abyss along elliptical orbits. His interest stems from the extra choices elliptical orbits seem to provide to mission-planning engineers. As he puts it: "The use of elliptic orbits allows much more flexibility for providing maximum coverage of specified geographic areas. Furthermore, the incredible crowding being experienced along the geosynchronous equatorial belt might be greatly alleviated by employing this new design approach."

The last two constellations we will consider in this chapter, Ellipso and Archimedes, employ elliptical orbits for their satellites whose elongated trajectories carry them repeatedly up through the Abyss. But, before we bring Ellipso and Archimedes onto the stage, we pause to review the special characteristics of Odyssey, a dozen-member constellation of satellites to be deployed in circular orbits 5600 nautical mi over the earth. Odyssey is being developed by TRW's talented engineers at Redondo Beach, California.

## TRW's Medium-Altitude Odyssey

After a series of careful computer simulations, TRW's experts quickly concluded that medium-altitude constellations appear to have a number of advantageous characteristics compared with competing constellations positioned at low-altitude or geosynchronous orbital locations.

In particular, medium-altitude constellations will probably turn out to be less expensive than either of the other two alternatives. Geosynchronous constellations give adequate coverage with fewer satellites, but each of those satellites is extremely expensive to build and transport into space. Low-altitude constellations can provide communication services with satellites that are cheaper to build and install, but extremely large numbers of them are required to achieve adequate global coverage.

As the discussions in Chapter 6 indicate, medium-altitude constellations also entail a number of other beneficial characteristics. These include

1. Infrequent and short-duration day-night cycling with the satellite shrouded in darkness only about 2 percent of the time

2. Moderate signal-propagation time delays

3. Slowly varying elevation angles with the satellites remaining high above the horizon most of the time

4. Small numbers of nearby space debris fragments menacing the satellites

5. Affordable possibilities for incremental startup with a partial constellation bringing in revenues from limited geographical areas to help pay for the unbuilt portions of the system

These and other similar study results have motivated TRW's managers and engineers to begin designing their Odyssey constellation. When Odyssey is complete, it will consist of 12 satellites orbiting the earth in medium-altitude orbits with inclinations of 55 degrees.

### Odyssey's Constellation Architecture

The master storyteller Homer, who probably resided in Greece around 700 or 800 B.C., concocted an exciting epic poem, *The Odyssey,* in which he recounted the adventures of Ulysses, who wandered up and down, around, and under the Mediterranean Sea for a stimulating decade immediately following the Trojan War.

During that 10 years of intermittent harassment and pain, interspersed with adventuresome events, Ulysses blinded the Cyclops and was captured and held prisoner on several different occasions. At one juncture he instructed his crew to rope him to the mast of his ship so he could listen to and enjoy the seductive song of the sirens without risking seduction followed by certain death.

Odyssey has been adopted as the colorful and exciting name for TRW's constellation of mobile communication satellites, which is sketched in Fig. 10.1. When it is complete, Odyssey will consist of twelve 4200-lb satellites launched into three orbital rings at an orbital altitude of 5600 nautical mi. Each orbit will be tipped 55 degrees with respect to the earth's equator.

As an Odyssey satellite travels around its orbit, it systematically swivels to compensate for its forward motion. Thus, its footprint covers the same region on the ground for an extended interval. This approach toward precise and accurate attitude control minimizes the number of hand-offs, which in turn helps

---

## ENJOYING ARMCHAIR VACATIONS THROUGH THE MAGIC OF INTERACTIVE SATELLITES

The special two-way interactive capabilities of spaceborne relays may soon allow you to go on an exciting vacation without ever leaving the comfort of your easy chair.

Would you like to spend a lazy afternoon poking around the ancient ruins of Pompeii? When the right satellites are all in place, you can begin your "armchair vacation" by picking up your videophone and dialing a proxy robot service that will put you in the "driver's seat." Once you have established the proper link for interactive communication, you can rent an Italian robot for an hour, a day, a week—whatever you may prefer. Sitting at the control panel in your living room, you begin maneuvering your feisty little robot in Naples—the closest city to Pompeii with proxy robots available for rent!

Soon you are urging your robot out onto the sidewalk where she mingles with the other tourists passing by. Pompeii is a bus ride away, so you guide her to the nearest bus stop while watching for obstacles with her television eyes. Soon the bus arrives, and you pay her fare with the credit card chained around her neck.

Along the way, you chat with the other tourists and buy a few souvenirs, which your robot takes to a packing and posting service. You also take color photographs with the camera thoughtfully provided by the hiring service where you picked her up. Those souvenir photos can be mailed to your home or digitally relayed back through interactive satellites.

Of course, you will not have the chance to enjoy the local cuisine—or sample those tasty Italian wines. But, by way of compensation, you will not likely suffer the ravages of food contamination. Muggings, mosquito bites, and fatal accidents will also be conveniently avoided during your armchair vacation.

So how about setting up another trip? Would you like to try a relaxing robot-activated climb up the Matterhorn? A chilly picnic set up on an Antarctic glacier? Or a leisurely afternoon rummaging through the Titanic's rusting hulk? Still not satisfied? How about relishing the beauties of the Amazon rain forest as seen through the eyes of a remotely piloted flying squirrel?

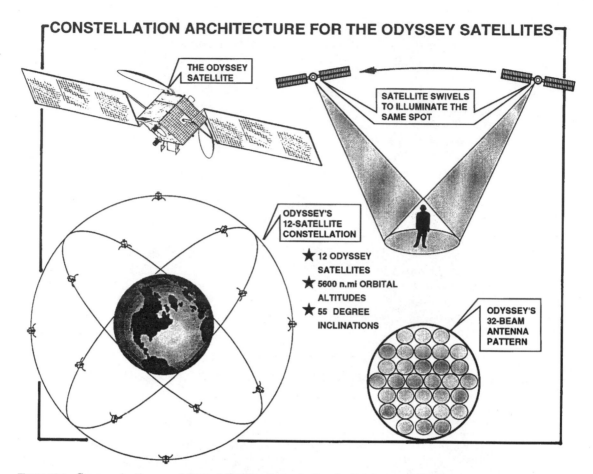

**Figure 10.1**  Communication specialists at TRW in Redondo Beach, California plan to provide mobile telecommunication services to a worldwide clientele using a tailor-made constellation of 12 Odyssey satellites 5600 nautical mi above the earth. Each 4200-lb Odyssey satellite will send down 32 spot-beams using frequency-efficient code division multiple access modulation techniques. As each satellite sweeps across the sky, it swivels to compensate for its forward motion. This dynamic attitude control approach allows the satellite to service a fixed region on the ground below for an extended interval, thus minimizing the number of hand-offs required.

maintain seamless service with a minimum number of drop-outs for the ground-based users. The Odyssey satellites employ CDMA modulation techniques with each satellite sending down 32 spot-beams to help ensure effective frequency reuse for enhanced spectrum utilization.

Single-satellite coverage could be provided with as few as nine Odyssey satellites. But TRW's larger 12-satellite constellation provides dual-satellite coverage for more robust user services. The extra satellites, in particular, provide a choice of two or more bent-pipe transmission paths, so the satellite chosen is usually high above the horizon. This helps minimize blockages due to vegetation, trees, buildings, and other structures. New calls are typically assigned to rising satellites. As a satellite descends over a particular region, calls from that region are gradually shed.

In addition to voice and data messaging, the Odyssey constellation also provides radionavigation services to selected users. The CDMA modulation techniques with their high chipping rates provide enhanced accuracy for the radionavigation solutions.

## Spacecraft Design Techniques

The 4200-lb Odyssey satellites are simple, nonprocessing "bent-pipe" transponders that employ electronic matrix amplifiers to deliver 3000 high-quality voice circuits per satellite focused primarily on the high-demand population centers in the industrialized countries of the world. Two solar arrays mounted on opposite sides on each spacecraft deliver 3100 watts of end-of-life electrical power.

Satellites in low-altitude orbits appear to race across the sky at 10 to 20 degrees/min, but the Odyssey satellites in their medium-altitude orbits move only 1 degree/min—about the width of a single finger held at arm's length. This slow, lumbering satellite movement helps minimize hand-offs to new satellites, thus cutting the risk of disconnects—which tend to occur during hand-offs for both terrestrial and spaceborne mobile communication systems.

Each satellite (Fig. 10.2) transmits 32 spot-beams down toward earth. Hybrid networks driven by matrix amplifiers allow dynamic (real-time) power redistribution for enhanced coverage for regions with large concentrations of paying clients. If desired, as much as 25 percent of the satellite's power can be routed into a single beam. This has an important impact on the services provided. As TRW's Roger J. Rusch enthusiastically explains, "By positioning and steering the satellite toward high-demand areas, up to 75 percent of its total capacity can be delivered to a region covering only 10 percent of the area being served."

The 5600-nautical-mi-high Odyssey satellites occupy the gap between the lower and the upper Van Allen Radiation Belts, but even so they are exposed to considerably more damaging radiation than any of their low-altitude or geosynchronous competitors. TRW's spacecraft designers are convinced that they can compensate for the extra radiation by choosing proper components, shielding them from damaging radiation, and building in sufficient redundancy to provide for a 10-year on-orbit mission life.

## Odyssey's Pocket Telephones

Odyssey's hand-held pocket telephones are being designed to radiate 0.5 watt of RF electrical power. They will provide full-duplex communication capabilities with batteries that can handle 90 minutes of talk-time and 24 hours of operation in the standby mode. As Fig. 10.3 indicates, the pocket telephone communicators will also feature alphanumeric messaging capabilities. Odyssey's uplink frequencies (user to satellite) occupy the L-band portion of the frequency spectrum between 1610 and 1625.5 MHz. Their downlink frequencies are in the S-band between 2483.5 and 2500 MHz.

The Odyssey communicators are being rigged for compatibility with existing terrestrial cellular telephone systems. When the user makes a call, the system

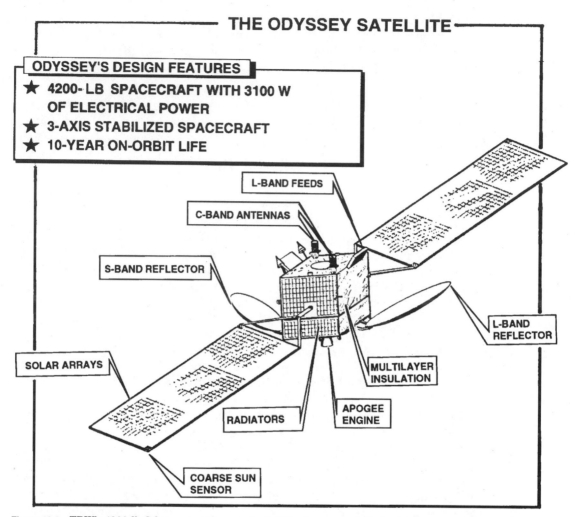

# THE ODYSSEY SATELLITE

**ODYSSEY'S DESIGN FEATURES**

★ **4200-LB SPACECRAFT WITH 3100 W OF ELECTRICAL POWER**
★ **3-AXIS STABILIZED SPACECRAFT**
★ **10-YEAR ON-ORBIT LIFE**

L-BAND FEEDS

C-BAND ANTENNAS

S-BAND REFLECTOR

L-BAND REFLECTOR

SOLAR ARRAYS

MULTILAYER INSULATION

RADIATORS

APOGEE ENGINE

COARSE SUN SENSOR

**Figure 10.2**  TRW's 4200-lb Odyssey is a three-axis stabilized spacecraft that uses yaw steering to enhance electrical power production while minimizing mechanical complexity. During its 10-year on-orbit life, each satellite generates an average of 3100 watts of electrical power. The 12 Odyssey satellites working in concert can provide mobile communication services for 2.3 million subscribers at a service rate of 65 cents for each minute during which they are connected.

automatically attempts to set up an inexpensive terrestrial cellular telephone relay link. If a terrestrial link is not available, the system defaults to a more expensive Odyssey satellite relay. Talk-time via the $1.3 billion Odyssey system is expected to cost about 65 cents per minute.

## Ellipso's Elliptical Orbit Approach

Mobile Communications Holdings, Inc. in Washington, D.C. is masterminding the Ellipso mobile communication system, which is being structured with a combination of elliptical orbits and circular orbits in the medium-altitude

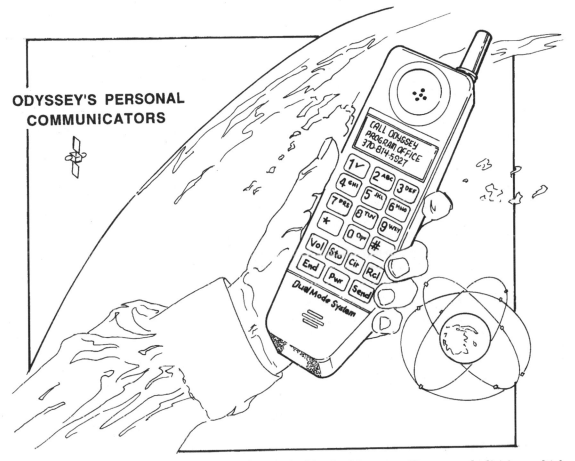

**Figure 10.3** The hand-held personal communicators serviced by the Odyssey satellites use code division multiple access modulations to serve a worldwide collection of mobile communication users. Odyssey is a dual-mode system. This means that it first attempts to set up a conventional terrestrial communication link before defaulting to a more expensive satellite link.

space flight regime. Ellipso's orbits are being selected to take advantage of the distribution of the land masses and the people who live on the surface of the earth.

## The Ellipso Constellation

The unique design for the Ellipso constellation came into being because the professional engineers who were assigned to design the Ellipso system noticed that the population densities in the northern and the southern hemispheres are distinctly different. The northern hemisphere, specifically, contains many heavily populated land masses above the 40 degree north latitude line. The southern hemisphere, by contrast, is virtually uninhabited below the 40 degree south latitude line.

Virtually all of Europe, for instance, half of the United States, and all of Canada are situated above 40 degrees north latitude. So are parts of Asia and northern Japan. In the southern hemisphere, by contrast, only relatively small portions of South America and New Zealand lie below 40 degrees south latitude. In short, almost all of the populated land masses on our planet lie above 40 degrees south latitude.

Ellipso's communications engineers set about to design the Ellipso constellation to match its capacity and resources with the earth's observed distribution of populated land masses. This they accomplished quite admirably by setting up essentially two separate, but coordinated, constellations of Ellipso satellites.

The largest of the two subconstellations will consist of as many as 15 elliptical-orbit satellites positioned in several different orbit planes with their apogees positioned over the northern hemisphere (Fig. 10.4). At apogee these satellites swing up to a 4212-nautical mi altitude. At perigee they dip down to an altitude of 281 nautical mi. All the elliptical orbits are inclined 116.5 degrees with respect to the equator. An elliptical orbit with that particular inclination does not rotate about its semimajor axis line. Consequently, its apogee will always remain over the northern hemisphere.

The apogees of the 15 Ellipso satellites launched into elliptical orbits are purposely positioned so they can serve mobile communication users in the densely populated regions of the northern hemisphere. When they are over the southern hemisphere, these Ellipso satellites are mostly quiescent. During quiescent intervals their batteries are being charged for use during peak-load intervals later on.

Ellipso's elliptical-orbit satellites are launched into sun-synchronous orbits with their semimajor axis lines carefully positioned so they will arrive over heavily populated regions of the northern hemisphere at specific times of the day. This flexible approach is possible only because the satellites are in elliptical orbits, which afford more degrees of freedom for mission adjustments than circular orbits.

The mid-latitudes of the earth, including the populated latitudes in the southern hemisphere, are covered by six additional satellites placed in circular equatorial orbits (Fig. 10.4). The altitudes for these circular orbits are 4212 nautical mi.

## Spacecraft Design

The Ellipso satellites are being designed for a 5-year on-orbit mission life. For design purposes, company engineers divided the visible portions of the earth into 19 adjacent zones each served by its own satellite beam. Each zone encompasses approximately the same surface area, but some beams cover greater distances than others. The beams that traverse the greatest free-space distance are rigged with extra power to compensate.

CDMA modulation techniques increase the communication capacity and efficiency of the Ellipso constellation. Each of the 19 beams carries the entire 16.5-MHz mobile satellite service band. This eliminates the necessity for

# THE ELLIPSO CONSTELLATION

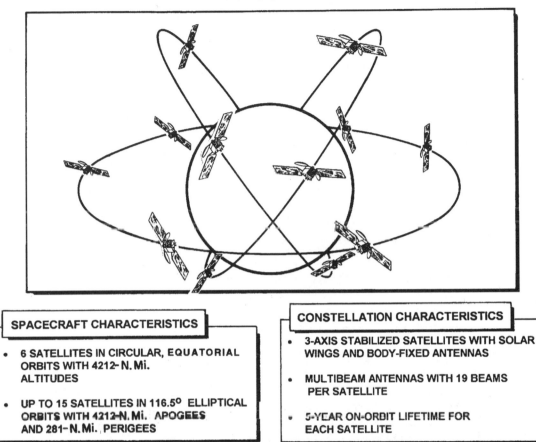

**SPACECRAFT CHARACTERISTICS**

- 6 SATELLITES IN CIRCULAR, EQUATORIAL ORBITS WITH 4212- N. Mi. ALTITUDES

- UP TO 15 SATELLITES IN 116.5° ELLIPTICAL ORBITS WITH 4212-N. Mi. APOGEES AND 281- N. Mi. PERIGEES

**CONSTELLATION CHARACTERISTICS**

- 3-AXIS STABILIZED SATELLITES WITH SOLAR WINGS AND BODY-FIXED ANTENNAS

- MULTIBEAM ANTENNAS WITH 19 BEAMS PER SATELLITE

- 5-YEAR ON-ORBIT LIFETIME FOR EACH SATELLITE

**Figure 10.4** Ellipso's aerospace engineers have designed their satellite constellation using a skillful combination of circular and elliptical orbits to concentrate the satellite's most intensive coverage where the largest numbers of people live. Six Ellipso satellites positioned in circular equatorial orbits 4212 nautical mi high cover a thick cummerbund surrounding the earth's equator. As many as 15 additional Ellipso satellites in inclined elliptical orbits with 4212-nautical mi apogees provide ample coverage for the thickly inhabited population centers in the northern hemisphere. The elliptical-orbit satellites are tipped 116.5 degrees with respect to the equator because an orbit with that particular inclination does not rotate about its semimajor axis line.

switching the communicator to a new frequency when it passes into a new beam, or when a new beam sweeps across its location. The Ellipso satellites are three-axis stabilized. They employ momentum wheels and thrusters for attitude control and any required stationkeeping maneuvers. Batteries power each satellite's transmitters during solar eclipse periods. Deployment of the first Ellipso satellites is presently scheduled for 1996.

The pocket telephone communicators will be similar in form and function to today's cellular telephones. Two primary types of communicators are being designed: (1) mobile or portable communicators and (2) hand-held communicators. The mobile units are being designed primarily for vehicular use. Like the smaller hand-held units, their antennas will provide hemispherical coverage patterns. Both types of communicators will be dual-mode devices designed for compatibility with terrestrial cellular telephone relays—when they are locally available.

Officials at Mobile Communications Holdings, Inc. have selected the Westinghouse Electric Corporation of Baltimore, Maryland to design and develop the ground infrastructure to serve their Ellipso constellation. Fairchild Space and Defense Corporation of Germantown, Maryland will build the satellite frames and the Harris Corporation of Melbourne, Florida will supply the communication payloads.

The Ellipso system is expected to cost a total of $700 million. Company officials believe they can break even with 350,000 subscribers. User charges will probably amount to $50 per month, plus 50 cents per minute for talk-time.

## Archimedes

Over the past 30 years, scientists and engineers in the former Soviet Union have launched nearly 150 Molniya communication satellites into inclined elliptical orbits serving high northern latitudes—with emphasis on the territories within the former Soviet Union. Each Molniya satellite typically weighs about 3500 lb. These satellites have been used mostly for telephone and television relay. But they have also been used for more exotic applications such as maintaining voice communications with manned space stations the Soviets have blasted into orbit around the earth.

Molniya satellites are placed in 12-hour orbits with their apogees situated above the northern hemisphere. They cross the equator at various points, but their orbital inclinations are always set at 63.4 degrees. That particular inclination (or the complimentary one at about 116.5 degrees!) keeps the satellite's apogee near the orbit's northernmost latitude point.

In 1990 European Space Agency (ESA) officials released a report indicating that they were planning to institute mobile communication services for the European continent using four satellites launched into 12-hour Molniya orbits. They decided to call their proposed constellation "Archimedes."

## Naming the Archimedes Constellation

Archimedes, an ancient mathematician and mechanical engineer, was born in Greece in 287 B.C. He made several important scientific discoveries including the physical principles of buoyancy and displacement. His device for moving water, called the Archimedes screw, has been in continuous use in irrigation ditches throughout the world for nearly 2300 years.

Archimedes was taking a bath when he suddenly figured out how to determine if King Hiero's crown was made of pure gold. This clever and insightful discovery excited Archimedes so much he raced through the streets of Athens in his birthday suit shouting "Eureka! Eureka!" (I have found it! I have found it!).

# WORLDWIDE WEATHER WATCH

Orbiting satellites help keep us safe. Sea-churning winds can still devastate wide swatches of terrain, especially along the Atlantic coast. But with the penetrating "eyes" of our weather satellites looking down from outer space, destructive hurricanes can no longer sneak ashore undetected—as did the "big wind" of 1938. Moreover, once a hurricane does strike, police, firemen, and paramedics can coordinate their rescue efforts using spaceborne communication links.

America's meteorologists had picked up reports of the hurricane of '38 when it was still paralleling the Atlantic coast. But they confidently predicted that it would soon "whirl harmlessly out into the Atlantic." Unfortunately, hurricanes do not always "whirl" as expected, and this one chose the worst imaginable spot to make a sharp left turn that carried its pounding winds—peaking as high as 186 mi/h—directly toward some of our country's heaviest population centers dimpling the shoreline between Boston and New York.

In the words of one contemporary account, "High tide rose 12 to 25 ft inundating the land and washing buildings out to sea." Wind-driven water, with buckshot fury, killed 680 people, seriously injured 700 more, and damaged $400 million worth of public and private property. It smashed 26,000 family cars, splintered more than 2000 private boats, and destroyed 15,000 buildings. And it flattened, snapped, or uprooted 275 million trees.

It also flattened America's faith in the science of meteorology. "A sophisticated population died by the thousands with little knowledge of what raw shape death took when it struck from the sky," wrote one contemporary observer. "In the long and laudable annals of the government's weather forecasts, that day's record makes what must be the sorriest page."

Nothing can stop the fury of a hurricane intent on destruction. But, in the age of space, some of its most devastating effects can be mitigated. Modern weather satellites can follow the course of a hurricane hour by hour, giving clear-cut warnings as to where it might strike next. And once it does strike, communication satellites, so sadly lacking in 1938, can maintain continuous and secure links between isolated victims and the rescue forces straining so hard to save lives.

Not without difficulty, Archimedes perfected a mathematical procedure that allowed him to estimate the value of pi to a high degree of accuracy. When he came to understand how the simple lever works, he is said to have remarked: "Give me a place to stand, and I will move the entire earth."

ESA has made an excellent choice in naming their constellation of mobile communication satellites Archimedes in honor of that joyous and insightful Greek mathematician.

## Constellation Architecture

The ascending nodes of the four Archimedes satellites are spaced 90 degrees apart as indicated by the sketch in Fig. 10.5. When each satellite travels up over the northern hemisphere, it provides communication relay services for a 6-hour interval. When it travels southward again, it encounters a 6-hour interval during which it is quiescent so its batteries can be recharged. During every other 6-hour active orbital segment, the Archimedes satellite is in the vicinity of Europe with an elevation angle above the local horizon of 70 degrees or greater. Consequently, obstruction on the ground is not a particularly serious problem.

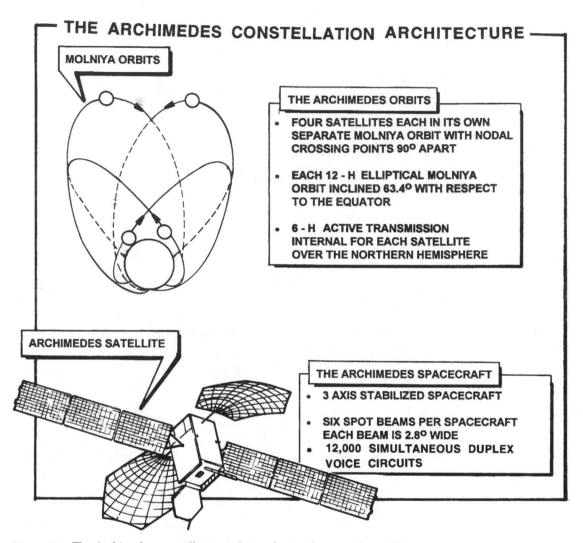

## THE ARCHIMEDES CONSTELLATION ARCHITECTURE

**MOLNIYA ORBITS**

**THE ARCHIMEDES ORBITS**

- FOUR SATELLITES EACH IN ITS OWN SEPARATE MOLNIYA ORBIT WITH NODAL CROSSING POINTS 90° APART

- EACH 12 - H ELLIPTICAL MOLNIYA ORBIT INCLINED 63.4° WITH RESPECT TO THE EQUATOR

- 6 - H ACTIVE TRANSMISSION INTERNAL FOR EACH SATELLITE OVER THE NORTHERN HEMISPHERE

**ARCHIMEDES SATELLITE**

**THE ARCHIMEDES SPACECRAFT**

- 3 AXIS STABILIZED SPACECRAFT

- SIX SPOT BEAMS PER SPACECRAFT EACH BEAM IS 2.8° WIDE
- 12,000 SIMULTANEOUS DUPLEX VOICE CIRCUITS

**Figure 10.5** The Archimedes constellation is being designed to provide mobile voice messaging services to the European continent using only four satellites placed in four separate inclined elliptical Molniya orbits. The ascending nodes of the four Archimedes satellites are spaced 90 degrees apart with their apogees positioned near their northernmost latitude points. Each Archimedes satellite, which orbits at a 63.4 degree inclination, travels around the earth twice every 24 hours. Then it starts all over again and traces out exactly the same route.

## System Design Characteristics

The Archimedes constellation architecture and the corresponding spacecraft design are being managed by the communication specialists at British Aerospace. Financial support for the system was authorized by ministers from ESA at a 1991 meeting in Munich, Germany. They are planning to launch their first Archimedes spacecraft in 1997. Eventually they may enlarge the constellation to provide global telecommunication coverage.

Archimedes is a high-power system (2500 watts) that can access inexpensive communicators at a low cost per channel. Frequency allocations are being pursued in the 1.5-GHz regime. Six spot-beams each 2.8 degrees wide will provide multinational coverage for the European continent and beyond. A 16.5-ft spacecraft antenna will provide the mobile transmission links. A total of 12,000 voice channels are being planned for the Archimedes system, 3000 from each of its four satellites.

# Summaries and Predictions

# 11

# Summing Up

*Everything that can be invented has been invented.*
*Letter received by President William McKinley from*
*the U.S. Patent Office, 1899*

The last four chapters of this book can be regarded as a dense tangle of details centering around 13 mobile communication constellations that differ from one another in a variety of interesting and important ways. The largest constellation includes at least 840 satellites skimming over the ground at a 378-nautical mi altitude. The smallest employs only one satellite hovering motionless at an altitude of 19,300 nautical mi. The simplest system relays only the briefest telegram messages. The most complicated provides voice messaging, geolocation, and even video imaging relay. The least expensive carries a total price tag of only $400 million. The most expensive will cost its sponsors at least $9 billion—or more than 20 times as much!

By now your head is probably crammed with a multitude of details from the previous chapters. So this chapter will, for purposes of clarification, sum up some of the more important characteristics of the various proposed constellations. The goal of these compact summaries is to give you a firmer grasp of today's mobile communication systems as compared with one another in all their rich and intricate detail.

## Constellation Location Comparisons

The last four chapters have included lengthy narrative descriptions of various mobile communication constellations positioned in medium-altitude, geosynchronous, and low-altitude orbital locations. In this section, and in the ones to follow, we will make various side-by-side performance comparisons between these proposed communication systems to see what kind of interesting patterns may begin to emerge.

## CONSTELLATIONS IN THE SKY

Ancient mariners spent many long, hypnotic hours studying and cataloging the formations of stars twinkling in the night sky. Those comforting constellations, together with the earth's magnetic field, helped them pick their way across vast uncharted seas. If they could have traveled, instead, across the sands of time, they would have been amazed to learn that we lackadaisically position artificial stars and talkative mechanisms along the celestial sphere.

In playful moments those ancient mariners identified their favorite constellations with the names of everyday animals (Pisces the fish, Taurus the bull) or mythological creatures (Perseus, Andromedia, Orion). We name our constellations after atoms (Iridium), ancient adventure stories (Odyssey), or tough old Greek mathematicians (Archimedes). Sometimes we commission advertising agencies to conjure up strange and cryptic names that sound weird to the human ear (Orbcomm, Globalstar, Glonass, Teledesic).

By the 12th century A.D., the magnetic compass was passing into widespread use among sophisticated maritime nations. One of the earliest references to compass navigation was published in 1188 when Englishman Alexander Neckam penned his colorful description of a primitive version consisting of "a needle placed upon a dart which sailors use to steer when the Bear is hidden by clouds."

Eighty years later, Dominican friar Vincent of Beauvais explained how masters of the sea whose ships were shrouded in dense fog would "magnetize the needle with a loadstone and place it through a straw floating in water." He then went on to note that, "when the needle comes to rest, it is pointing at the Pole Star."

In 1492, fretful sailors aboard the *Santa Maria* were rapidly careening toward revolt because, as they sailed further and further from the land they knew, their magnetic compass refused to seek out the North Star. Imagine how much easier that agonizing journey might have turned out to be if today's supportive constellations of navigation and communication satellites had been available to guide and comfort them as they crossed the open sea!

## The Medium-Altitude and the Geosynchronous Constellations

The orbital locations for five medium-altitude and geosynchronous constellations are depicted in Fig. 11.1. The constellation altitude is indicated by the radial coordinate, whereas its orbital inclination is indicated by the angular coordinate running along the figure's outer edge.

Notice how the three medium-altitude constellations—Elipso, Archimedes, and Odyssey—are all clustered together in the left-hand portion of this polar-coordinate graph. The two elongated cylinders are used to indicate elliptical-orbit constellations. The top end of the cylinder marks the apogee altitude of the satellites; the lower end marks their perigee location.

The Ellipso satellites are to be launched into a split constellation. Fifteen of the Ellipso satellites are destined for elliptical orbits with apogee altitudes of 4212 nautical mi tipped at inclination angles of 116.5 degrees. Six of the Ellipsos are in circular, equator orbits at 4212 nautical mi. The major benefit of splitting the Ellipso constellation is to provide ample low-cost coverage for essentially all of the heavily inhabited portions of the earth above 40 degrees south latitude.

The four Archimedes satellites are to be lofted into 12-hour elliptical Molniya orbits with 63.4 degree orbital inclinations. Notice how the greatly elongated

# MEDIUM-ALTITUDE & GEOSYNCHRONOUS MOBILE SATELLITE CONSTELLATIONS

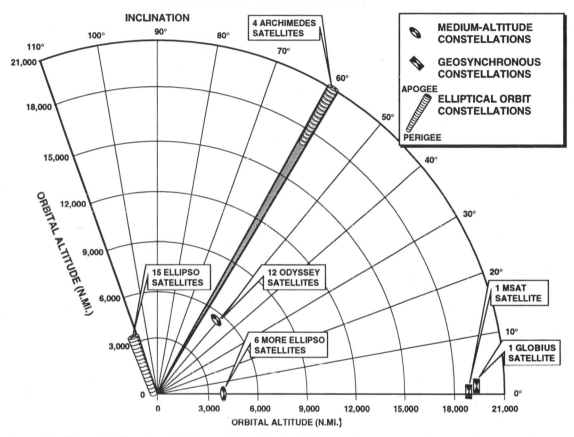

**Figure 11.1** The orbital locations for five medium-altitude and geosynchronous mobile communication constellations are depicted in this polar coordinate graph. The radial and the angular coordinates denote orbital altitude and inclination, respectively. The two tapered cylinders are used to represent elliptical-orbit constellations. The top end of each cylinder represents the constellation's apogee location; its lower end marks the perigee location. Notice how the two geosynchronous constellations are positioned over the equator. All the others involve satellites in highly inclined orbits.

Molniya orbits cause the Archimedes satellites to swing a little above the geosynchronous altitude when they travel up toward their apogees.

The American Mobile Satellite Corporation's (AMSC) MSAT and the Russian Globis orbital antenna farm are headed toward geostationary orbits 19,300 nautical mi high. Consequently, the rectangles that mark their orbital locations are positioned on the far right in the polar coordinate graph. An examination of Fig. 11.1 might seem to indicate that MSAT and Globis are squeezed up close together at geosync. However, the graph in Fig. 11.1 shows only altitude and inclination—longitude is not depicted at all. In the real-life situation, depending on the orbital slots actually selected, MSAT and Globis could end up on opposite sides of the earth, thousands of miles apart.

# SATELLITES POSITIONED IN LOW-ALTITUDE VOICE-MESSAGING AND TELEGRAPH CONSTELLATIONS

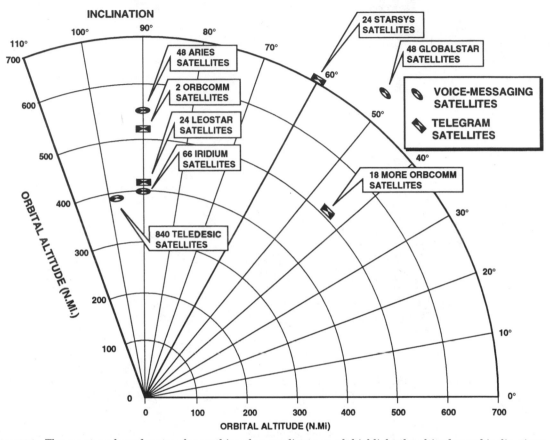

**Figure 11.2**  The squat ovals and rectangles on this polar coordinate graph highlight the altitudes and inclinations of various low-altitude voice messaging and telegram-style constellations. The sun-synchronous Teledesic constellation skims over the earth at the lowest altitude (378 nautical mi), but with the highest orbital inclination (98.2 degrees). Notice how Orbcomm is split into two subconstellations with 2 satellites occupying polar orbits and 18 more at the same 524-nautical mi altitude with inclinations of 45 degrees.

## The Low-Altitude Constellations

A similar polar coordinate graph in Fig. 11.2 pins down the orbital altitudes and inclinations for the various low-altitude constellations. In this case, the orbital positions of the more sophisticated voice messaging satellites are represented by ovals, and the simpler telegram-style satellites are represented by rectangles.

As Fig. 11.2 clearly indicates, the low-altitude constellations tend to have relatively high orbital inclinations. The choice of orbital inclination is driven primarily by the desire to achieve worldwide coverage from the constellation—or at least coverage for an appreciable fraction of the earth.

Orbcomm is a split constellation. Two of its satellites are to be positioned in 524-nautical mi polar orbits; the other 18 Orbcomms are targeted toward circular orbits with 45 degree inclinations at the same 524-nautical mi orbital altitude.

Teledesic is the largest and most ambitious constellation proposed so far with at least 840 satellites to be launched into 378-nautical mi sun-synchronous orbits. The Teledesic constellation is intended to provide fixed telephone services with 2 million duplex voice channels available to serve users worldwide.

## The Salient Features of Various Mobile Communication Systems

Table 11.1 summarizes some of the more important features designed into the medium-altitude and the geosynchronous constellations. Notice how entries in the table focus on the various constellation geometries, the services each constellation provides, the electromagnetic frequencies used by its communication links, and a sampling of some of its more important spacecraft characteristics. Similar comparisons are provided in Table 11.2 for the low-altitude constellations. The tabular entries in the top half of the table characterize the low-altitude voice messaging constellations. Those in the bottom half of the table apply to the simpler telegram-style constellations.

## Uplink and Downlink Frequency Selections

The uplink and downlink frequency selections for the medium-altitude and the geosynchronous constellations are highlighted in Fig. 11.3. L-band frequencies in the vicinity of 1.5 to 1.6 GHz are the selection of choice for the medium-altitude voice messaging constellations—Odyssey, Ellipso, and Archimedes. The MSAT voice messaging system at the geosynchronous altitude also employs L-band frequencies for the same purpose.

The 42,000-lb Globis orbital antenna farm, by contrast, is slated to operate in several different frequency bands ranging from 1 to 18 GHz. Russian marketers who are touting the Globis are hoping to persuade international regulatory agencies to grant them access to a variety of otherwise underutilized frequency bands.

The uplink and downlink frequencies for the low-altitude voice messaging and low-altitude telegraph satellites are presented in Fig. 11.4. Generally speaking, the low-altitude voice messaging constellations operate in or near the L-band and S-band frequencies between 1.5 and 2.5 GHz.

The Teledesic constellation is a major exception. In order to obtain ample bandwidth for their proposed market of 20 million customers, Teledesic engineers are planning to operate in the Ka-band between 20 and 30 GHz. Rain attenuation can be a worrisome problem at the extremely short Ka-band wavelengths, but the Teledesic engineers hope to alleviate that problem via path diversity and relatively high elevation angles provided by their large constellation of satellites.

TABLE 11.1  The Salient Features of the Medium-Altitude and the Geosynchronous Mobile Satellite Constellations)

| CONSTELLATION | CONSTELLATION GEOMETRY | SERVICES PROVIDED | SPACECRAFT CHARACTERISTICS | USER FREQUENCIES | COST ESTIMATES |
|---|---|---|---|---|---|
| **THE MEDIUM-ALTITUDE CONSTELLATIONS** | | | | | |
| ODYSSEY Sponsored by TRW, Inc. in Redondo Beach, California | 12 Satellites in 5600 n.mi. circular orbits with 55° inclinations in 3 orbit planes | Mobile voice, data services, and geolocation 23,000 simultaneous voice channels. Geolocation accuracy = 1300 feet. | 4200-pound satellites drawing 3100 watts over a 10-year on-orbit life. 3-axis stabilization. 32 downlink beams per satellite. | UPLINKS 1610 to 1626.5 MHz  DOWNLINKS 2483.5 to 2500 MHz | Total System cost = $1.3 billion Communicator cost = $300 (ultimately) Cost per call = 65¢ per minute |
| ELLIPSO Sponsored by Mobile Communications Holdings, Inc. in Washington, D.C. | Up to 15 satellites in elliptical orbits (281 by 4,212 n.mi.) with 116.5° inclinations plus 6 satellites in circular, equatorial orbits at 4,212 n.mi. altitudes | Mobile voice, data, fax, and computer services. 350,000 subscribers needed to break even | Not Available | UPLINKS 1610 to 1616.5 MHz  DOWNLINKS 2483.5 to 2500 MHz | Total System cost = $700 million User charges = $50 per month plus 50¢ per minute |
| ARCHIMEDES Sponsored by the European Space Agency | 4 Satellites in 12-hour molniya orbits with 90° spacings between their ascending nodes. Inclination angles = 63.4° | Mobile voice, data, fax, and computer services for the European continent. 12,000 simultaneous voice channels. | 3-axis stabilized satellites drawing 2500 watts. Six spot beams per satellite. 2.8° wide 16.5 foot spacecraft antenna for mobile services. | 1.5 gigahertz operating regime for uplinks and downlinks | Not Available |
| **THE GEOSYNCHRONOUS CONSTELLATIONS** | | | | | |
| MSAT Sponsored by American Mobile Satellite Corp. in Reston, Virginia | 1 Satellite in a circular equatorial geosynchronous orbit. An identical satellite is to be orbited by Bell Canada | North American coverage for carphones and fixed phones in unserved rural areas. Present authorization is for 2000 full duplex circuits | 6300-pound 3-axis stabilized spacecraft drawing 3000 watts. Spacecraft sends down 6 spot beams in the L-band. | UPLINKS 1631.5 to 1660.05 MHz  DOWNLINKS 1530 to 1559 MHz | Total System cost = $550 million Communicator cost = $2000 Cost per call = $1.45 (standard service) |
| GLOBIS Aerospace Consortium to be formed in the former Soviet Union with other world partners | 1 Satellite in a circular equatorial geosynchronous orbit. | 1400 duplex circuits for pocket telephones plus 300,000 duplex circuits for fixed ground stations and 30 channels of color T.V. | 12,000-pound 3-axis stabilized spacecraft drawing 15,000 watts. 30 spacecraft antennas plus 2 gigantic solar arrays. 10-year on-orbit life. | Several frequency bands ranging from 1 to 18 gigahertz | Not Available |

TABLE 11.2    The Salient Features of the Low-Altitude Voice Messaging and the Telegraph-Style Constellations

**LOW-ALTITUDE VOICE-MESSAGING CONSTELLATIONS**

| CONSTELLATION | CONSTELLATION GEOMETRY | SERVICES PROVIDED | SPACECRAFT CHARACTERISTICS | USER FREQUENCIES | COST ESTIMATES |
|---|---|---|---|---|---|
| IRIDIUM Sponsored by Motorola in Schaumburg, Illinois | 66 Satellites in 413 n.mi. circular orbits with 90° inclinations in 6 orbit planes | Digital voice, data services and geolocation on a global basis. Geolocation accuracy = 1 nautical mile | 1600-pound satellites drawing 1400 watts average power. Onboard switching & satellite-to-satellite crosslinks. 5 year on-orbit life. | UPLINKS 1610 to 1626.5 MHz  DOWNLINKS 1610 to 1626.5 MHz  CROSSLINKS 1610 to 1626.5 MHz | Total System cost = $3.4 billion Communicator cost = $2500 to $3000 cost per call = $2 to $3 per minute |
| GLOBALSTAR Sponsored by Loral Cellular (Loral/Qualcomm in New York City and San Diego, CA.) | 48 Satellites in 750 n.mi. circular orbits with 52° inclinations in 8 orbit planes | Prime coverage with digital voice, data and geolocation between ±70° latitude. 28,600 duplex and data channels. Geolocation accuracy = 1000 feet. | 490-pound satellites drawing 150 watts average power over a 7.5 year on-orbit life. | UPLINKS 1610 to 1626.5 MHz  DOWNLINKS 1610 to 1626.5 MHz and 2483.5 to 2500 MHz | Total System cost = $1.7 billion Communicator cost = $750 (ultimately) cost per call = 30¢ per minute |
| ARIES Sponsored by Constellation Communications, Inc. in Herndon, Virginia | 48 Satellites in 550 n.mi. circular orbits with 90° inclinations in 4 orbit planes | Digital voice, data and geolocation services worldwide. Geolocation accuracy = 3 or 4 nautical miles. | 275-pound satellites drawing 107 watts average power over a 5 year on-orbit life. | UPLINKS 1610 to 1626.5 MHz  DOWNLINKS 2483.5 to 2500 MHz | Total System cost = $400 to $500 million Cost per call = 30¢ per minute |
| TELEDESIC Sponsored by Teledesic in Kirkland, Washington | 840 Satellites in 378 n.mi. sun-synchronous orbits with 98.2° inclinations in 21 orbit planes | 2 million simultaneous full-duplex voice circuits (with video relay too). Primary market devoted to fixed telephone services. Video services also to be provided. | 1750-pound satellites rigged with swiveling multibeam antennas directed toward "stationary" super cells on the ground. | 20 to 30 gigahertz portion of the Ka frequency band (400 MHz of bandwidth needed) | Total System cost = $9 billion Communicator cost = $2000 then declining Cost per call = 25¢ to 30¢ per minute |

**TABLE 11.2** The Salient Features of the Low-Altitude Voice Messaging and the Telegraph-Style Constellations (Continued)

| CONSTELLATION | CONSTELLATION GEOMETRY | SERVICES PROVIDED | SPACECRAFT CHARACTERISTICS | USER FREQUENCIES | COST ESTIMATES |
|---|---|---|---|---|---|
| **LOW-ALTITUDE VOICE-MESSAGING CONSTELLATIONS** | | | | | |
| ORBCOMM Sponsored by Orbital Sciences Corporation in Vienna, Virginia | 18 Satellites in 424 n.mi. circular orbits with 45° inclinations in 3 orbit planes, plus 2 Satellites in 424 n.mi. polar orbits | Store and forward digital data communications and geolocation services positioning accuracy = 1200 feet. | 85-pound microstar satellites drawing 160 watts average power. Satellites launched in 8-packs: 41 inches in diameter and 6.5 inches high | UPLINKS 148 to 150.05 MHz   DOWNLINKS 137.2 to 138 MHz | Total System cost = $400 to $500 million Communicator cost = $100 to $400 |
| STARSYS Sponsored by Starsys Positioning, Inc., in Lanham, Massachusetts | 24 Satellites in 702 n.mi. circular orbits with 60° inclinations in 3 (or 4) orbit planes | Store and forward digital data communications and messaging. Positioning accuracy = 3300 feet | 220-pound satellites drawing 120 watts of power. 5 year on-orbit life. | UPLINKS 148 to 149.9 MHz   DOWNLINKS 400 MHz | Not Available |
| LEOSTAR Sponsored by Italspuzio in Rome, Italy | 24 Satellites in 432 n.mi. circular orbits with 90° inclinations in 4 orbit planes | Real time and store and forward communication services | 90 to 130-pound satellites drawing 100 to 200 watts of power stabilization via gravity gradient booms. | UPLINKS 950 to 959 MHz and 960 to 980 MHz   DOWNLINKS 905 to 914 MHz and 1000 to 1020 MHz | Not Available |

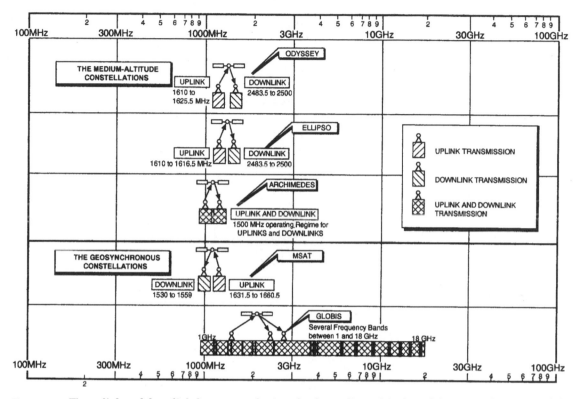

**UPLINK AND DOWNLINK FREQUENCIES FOR THE MEDIUM-ALTITUDE AND THE GEOSYNCHRONOUS MOBILE SATELLITE CONSTELLATIONS**

**Figure 11.3**   The uplink and downlink frequency selections for the medium-altitude and the geosynchronous mobile communication constellations are properly positioned on this logarithmic graph. Notice how mission planners for Odyssey, Ellipso, Archimedes, and MSAT are all planning to operate in the L-band portion of the frequency spectrum in the vicinity of 1.5–1.6 GHz. The frequencies for the Russian Globis, by contrast, will range from 1 to 18 GHz. Russia's engineers are hoping to negotiate successfully for the use of several narrow bands within that broad-ranging span.

The frequency selections for the low-altitude telegraph satellites are depicted along the bottom half of Fig. 11.4. All of their transmission frequencies are below 1020 MHz in special portions of the frequency spectrum that have already been set aside for the telegraph-style satellites.

## Cost Estimates

The bar charts in Fig. 11.5 represent the estimated system costs and the number of satellites for various mobile satellite voice messaging systems. Most of the cost estimates presented in this figure were supplied by the consortiums and private companies proposing the constellations to the FCC. The others came from various alternate sources. Whenever these sources differed with one another, a judicious selection was made from the best available information.

## UPLINK AND DOWNLINK FREQUENCIES FOR THE LOW-ALTITUDE VOICE-MESSAGING AND TELEGRAPH CONSTELLATIONS

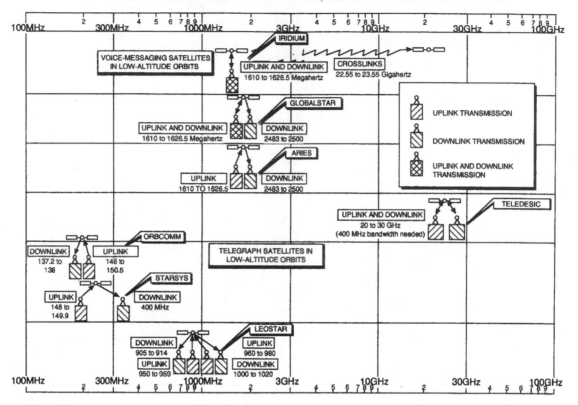

**Figure 11.4** The uplink and downlink frequencies for various low-altitude voice messaging satellite constellations are highlighted on this logarithmic graph. The low-altitude Iridium and Globalstar voice messaging systems share the L-band frequencies around 1.5–1.6 GHz with various higher-altitude voice messaging systems (see Fig. 11.3). Teledesic engineers are planning to operate their giant swarm of satellites in the 20- to 30-GHz regime where ample bandwidths could turn out to be available. The low-altitude telegram constellations do not require large bandwidths, so they can all operate below 1020 MHz in narrow slices of the frequency spectrum.

The total estimated cost for the least expensive voice messaging systems, Aries and MSAT, amounts to about a half-billion dollars each. Aries is a low-altitude constellation containing 48 satellites. MSAT contains a single geosynchronous satellite being financed by AMSC in Reston, Virginia. The most expensive system is, of course, Teledesic, which is being backed by American billionaires Bill Gates and Craig McCaw. Teledesic carries an estimated price tag of $9 billion. Thus, it is slated to cost about $1 billion more than the Alaskan Pipeline—which was also financed with private funds.

The other six mobile voice messaging systems depicted in Fig. 11.5 lie somewhere between these two cost extremes. Teledesic aside, Iridium is slated to be the most costly system at $3.4 billion. To put these various proposed expenditures into perspective, it may be helpful to note that they constitute an appreciable fraction of NASA's annual budget, which currently amounts to about $15 billion.

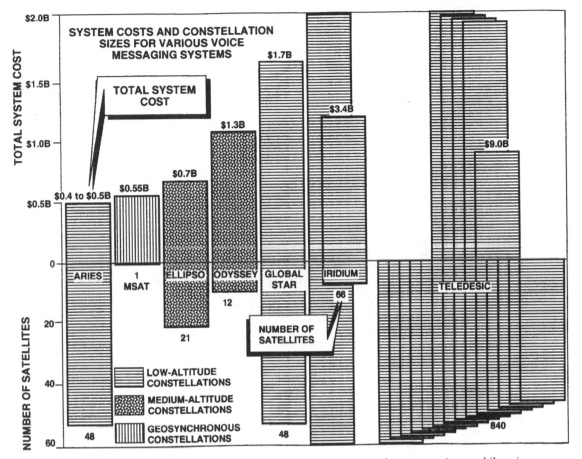

**Figure 11.5** These bar charts provide convenient side-by-side comparisons between various mobile voice messaging systems based in space. Aries, with 48 low-altitude satellites, and MSAT, with a single geosynchronous satellite, are expected to cost about a half-billion dollars each. Teledesic, with at least 840 satellites, is slated to cost about $9 billion. Dollarwise, building and installing the Teledesic constellation is thus roughly equivalent to constructing the Alaskan Pipeline all over again.

# SOME SPECTACULAR SPACE-AGE REPAIRS

Houston, we have a problem.
Say again, *Apollo 13*.
We have a problem.

It was a problem all right! Seconds before, a violent explosion had ripped through the Apollo Service Module, knocking out two of its three fuel cells and dumping the astronauts' precious oxygen supplies into black space.

At first they managed to remain fairly calm, but as their crippled spacecraft hurtled on toward the moon, a fresh crisis suddenly unfolded: The lithium hydroxide canisters in the LEM (Lunar Excursion Module) and the Service Module turned out to be noninterchangeable, and as a result, the air the astronauts were breathing was rapidly becoming polluted. Fortunately, they were able to patch together a workable connection to the canisters in the Service Module, thus making them usable in their overcrowded "lifeboat" LEM.

During the next few years other astronauts successfully achieved a number of other spectacular spaceborne repairs, thus proving that astronauts were definitely not merely along for the ride or "Spam in a can" as a cynical journalist once wryly observed. When the micrometeoroid shield was ripped off the main body of the *Skylab,* for instance, the astronauts erected a big cooling parasol to shield themselves from the burning rays of the sun. On the next mission, astronauts Jack R. Lousma and Owen K. Garriott remodeled the *Skylab's* parasol sunshade by erecting two 55-ft metal poles to form a large A-frame tent over their freshly occupied home in space. Other *Skylab* astronauts repaired an ailing battery, retrieved exposed film from the *Apollo* telescope mount, and removed and replaced several gyroscopes used in stabilizing their wobbling craft. These complicated tasks were all performed in full space suits outside the protective envelope of the *Skylab* modules.

The retrieval and redeployment of the *Solar Max* satellite—which was filmed with IMAX cameras operated by other space shuttle astronauts—provides another powerful illustration of the skill and dexterity of humans in space.

Space-age robots have also performed in a similarly impressive manner. For instance, when the television camera mounted on the elbow of the shuttle's 50-ft robot arm sent back pictures of a big chunk of ice growing on the outside of the waste-water vent on the shuttle orbiter, the Canadian robot arm helped the astronauts execute a clever solution. Rather than risk possible damage to the shuttle's delicate heat shield, should chunks of the ice break loose during reentry, the astronauts were instructed to use the robot arm like a big, heavy trip hammer to knock the ice loose.

On another mission, the robot arm was ready to release the Earth Radiation Budget Satellite into the blackness of space. Unfortunately, during deployment, its solar arrays got stuck in an awkward position, so the astronauts used the robot arm to shake the satellite vigorously. Then they held it up to the warming rays of the sun so its solar array could unfold.

# Crystal Ball Predictions for Century 21

*We must not be misled to our own detriment to assume that the untried machine can displace the proved and tried horse.*

MAJOR GENERAL JOHN K. HERR, 1938

This final chapter includes a few selected predictions concerning the future of spaceborne mobile communication systems. Crystal balls have never been very reliable guides to future courses of action, so these courageous predictions may turn out to be wildly inaccurate. But, nevertheless, if we are to plan for the future, we definitely need to formulate our best guess as to what that future might turn out to be.

"Making predictions is always difficult," observed the noted physicist, Hans Bethe, "especially when they involve the future." During his professional career, Bethe came to realize that, in the field of professional prognostication, there are few reliable guidelines for success.

Tea leaves, chicken entrails, the constellations of stars sprinkled across the celestial sphere have all been pressed into service in attempting to divine the future. But, even when shrewd prophets have been equipped with the finest available aids toward successful prognostication, their results have usually turned out to be unexceptional, at best. Botched prophecies have occasionally brought violent death. But, more often, the soothsayer has merely been subjected to stinging dollops of ridicule.

Two thousand years ago, Cicero was entirely unimpressed with the inaccurate predictions being offered up by the best-known prophets of his day. On one occasion he made a sarcastic address to the Roman senate: "It seems to me," he said, "that no soothsayer should be able to look at another soothsayer without laughing."

## CRYSTAL BALL PREDICTIONS THAT MISSED THE MARK

When technological advancements hurtle forward at a breakneck pace, even world-renowned experts have a difficult time seeing what lies ahead—even within their areas of expertise! Consider these embarrassing examples of expert opinion sadly missing the mark:

*There is no likelihood that man can ever tap the power of the atom.*

NOBEL PRIZE WINNING PHYSICIST
ROBERT MILIKEN, 1923

*There will probably be a mass market for no more than a thousand motorcars in Europe. There is, after all, a limit to the number of chauffeurs who could be found to drive them.*

SPOKESPERSON FOR DAIMLER-BENZ
CIRCA 1900

*We have struck an iceberg, but there is no danger.*

CAPTAIN'S ANNOUNCEMENT TO THE PASSENGERS
OF THE *TITANIC*, 1912

*He will never amount to anything.*

HIGH SCHOOL REPORT FROM MUNICH,
GERMANY, ATTEMPTING TO PREDICT THE
FUTURE ACCOMPLISHMENTS OF THE YOUNG
ALBERT EINSTEIN, 1892

*Atomic energy might be as good as our present-day explosives, but it is unlikely to produce anything very much more dangerous.*

FUTURE PRIME MINISTER OF ENGLAND
WINSTON CHURCHILL, 1939

*My only worry about Cassius Clay is whether they can get a doctor who can get my glove out of his mouth without cutting my wrist.*

BOXING CHAMPION SONNEY LISTON, SPEAKING
TO A GROUP OF NEWSPAPER REPORTERS THREE
DAYS BEFORE CASSIUS CLAY KNOCKED HIM OUT
IN THE THIRD ROUND, 1963

*Can't act. Can't sing. Slightly bald. Can dance a little.*

STUDIO REPORT EVALUATING FRED ASTAIRE'S
FIRST SCREEN TEST, 1932

Playwright Eugene Ionesco once gave some marvelous advice to ambitious prognosticators anxious to find fresh new ways to ply their risky trade. "You can predict things only after they've happened," he pointed out with a knowing grin.

## The Ultimate Personal Computer

In 1979, I wrote an account detailing the capabilities of the "Ultimate Personal Computer" in my book, *Our Computerized Society with Basic Programming* (Anaheim Publishing Company). My Ultimate Personal Computer (Fig. 12.1) had, for its day, some rather amazing capabilities. For instance, when you operate it in different modes, the functioning of the buttons *and* their electronic labeling change automatically!

The date is November 12, 1999. Wayne Bishop, a San Diego college professor, has just entered his bachelor apartment on the southeastern corner of Mission Bay. In his pocket, he carries a flat, battery-powered device covered with

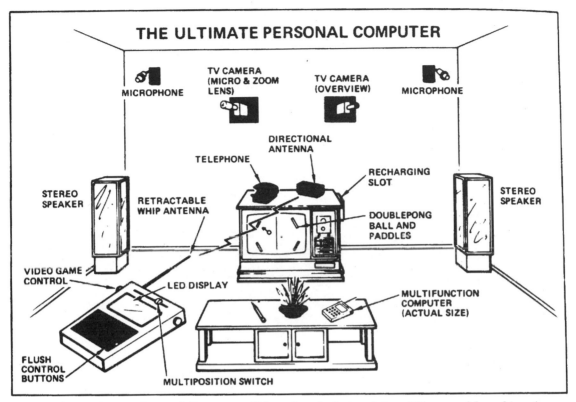

**THE ULTIMATE PERSONAL COMPUTER**

MICROPHONE

TV CAMERA (MICRO & ZOOM LENS)

TV CAMERA (OVERVIEW)

MICROPHONE

DIRECTIONAL ANTENNA

TELEPHONE

RECHARGING SLOT

STEREO SPEAKER

RETRACTABLE WHIP ANTENNA

DOUBLEPONG BALL AND PADDLES

STEREO SPEAKER

VIDEO GAME CONTROL

LED DISPLAY

MULTIFUNCTION COMPUTER (ACTUAL SIZE)

FLUSH CONTROL BUTTONS

MULTIPOSITION SWITCH

**Figure 12.1** The Ultimate Personal Computer will be a cleverly designed, incredibly powerful machine. Among other things, it will allow you to play video games with distant friends and give you instant access to distant data files. Constructed with flush control buttons and three rows of liquid crystal displays, the unit includes a special master control switch that allows you to change the function (and the labeling) of the other buttons and displays. A compact bubble memory handles the computer's scratchpad storage, remembers the telephone numbers of selected friends, and performs the functions of an electronic appointment book.

flush control buttons and three rows of liquid crystal displays. The compact little unit, which is sketched in Fig. 12.1, is a portable computer with some novel and impressive capabilities in its own right. But it is also an ingenious electronic key that unlocks larger computer facilities and remote data banks.

Wayne kicks off his shoes and drapes his tie over the back of an easy chair. Is there anything worthwhile on television in the next hour or two? After sliding the master control switch on his Ultimate Personal Computer to the proper position, he pushes a few buttons and his television set quietly springs to life. Listings of all the local, cable, and network shows flash on the screen. A play by Tennessee Williams is scheduled to begin a quarter of an hour later. The play looks promising, so he pushes a few more buttons. The telephone responds by connecting him to a special preview service which displays sample scenes on his television screen. Unfortunately, the play does not seem as interesting as he had hoped, so he flips the master control switch to a different position. In an instant, the buttons are labeled with familiar names. He pushes the one labeled "Marcia," and his telephone automatically connects him to his sister Marcia, who lives in New York City. (Her phone number is stored in the computer's bubble memory.) He keys his device to print a challenge on her television screen:

```
THIS IS THE RED BARON CALLING. OFFICIALLY
CHALLENGING YOU TO A GAME OF DOUBLEPONG.
```

Moments later, a curt reply races across his screen:

```
HI, WAYNE. GLAD TO HEAR FROM YOU.
WILL GLADLY ACCEPT YOUR CHALLENGE.
```

Wayne grasps his computer unit firmly with both hands. The spring-loaded controls for video games are located along the outer edges (Fig. 12.1). Subtle pressures guide the movements of his two paddles. He also sees her paddles zipping back and forth across his screen. His kid sister is surprisingly good at Doublepong! Halfway through the second game, he realizes that she is too much for him.

"O.K. I CONCEDE" he types on his unit. A quick message comes back on his screen. "I'VE BEEN PRACTICING. DO YOU HAVE THE TIME TO TALK?" He types "YES."

Then he presses a button to set up a voice link. Her voice comes over the stereo speakers loud and clear. They exchange verbal greetings and a little random smalltalk, then he asks her about the new model train she is assembling, complete with fake steam.

"It sounds terrific, Sis."

"You wanna see?" she asks him earnestly.

"Sure I do."

He presses another button and her full-color image fills the screen. She holds the engine in the palm of her hand. "It's beautiful," he tells her softly. She beams with pride. He wonders how the motor is constructed so Marcia adjusts her micro lens until the image of the tiny motor is 18 in. high. Satisfied with what he has seen, Wayne disconnects the video image. (Talking takes only 1/600th the bandwidth so it's much cheaper.)

She reminds him of the family reunion they have been planning for the end of the next month. He interrogates his computer to make sure he has not scheduled anything else for those crucial dates. Good, no conflicts! Then, to make sure he won't forget, he types in the December 8 reunion date and rigs the device to remind him on the 5th and on the 6th.

When Marcia rings off, he keys in the number of "airline central." A message flashes across the middle of his television screen:

```
AIRLINE CENTRAL. GOOD AFTERNOON.
WHAT CAN I DO FOR YOU?
```

He keys in his request.

```
DEC. 7 FLIGHT SCHEDULES.
SAN DIEGO TO NEW YORK.
```

The printed schedules rapidly appear. He selects one and keys in a firm request.

```
DEC. 7. RESERVATION CONFIRMED
8:45 A.M. AMERICAN AIRLINES FLIGHT 78, GATE 23
$500 DEDUCTED FROM YOUR ACCOUNT.
```

He then transfers the departure time and the scheduled date into his unit and instructs it to issue reminders 2 days and 1 day in advance. Then he begins to wonder if there is enough money in his account to pay for the tickets without going into the red. So he dials the bank's number and requests his balance. A computer's stiff voice comes back over the line. "VOICE PRINT PLEASE." After stating his name and his account number three times in rapid succession, the computer replies,

"Voice print analysis positive. Bank balance $955.39."

At the same time, it displays his current balance on the television screen so there can be no mistake. Wayne does a few calculations on his unit to make sure he won't have an overdraft before the end of the month. It looks like his finances will be O.K.

"How long has it been since I recharged the computer?" he wonders aloud. But he cannot remember for sure. Of course, the computerized unit will automatically remind him when its charge begins to fade; however, it's always better to play it safe. People have begun to depend on these versatile little devices so much they get terribly upset whenever they fail or run down their batteries. Wayne once encountered a young fellow in a downtown cocktail bar who was drowning his problems in 80 proof Scotch. It turned out that he was afraid to go home and face his family after he had lost their only pocket computer. Wayne punches two buttons and the display springs to life. "CHARGE REMAINING 1.2 HOURS." He walks over to his television set and drops his unit into a slot. A soft red light indicates that its batteries are picking up the charge. Within 2 hours, his Ultimate Personal Computer will be back up to full strength.

**Prediction:**   The Ultimate Personal Computer, a device that seemed so innovative and futuristic in 1979, will soon be in the hands of ordinary high school students. They will find it hard to believe that such machines did not always exist in every previous era.

## Keeping in Constant Touch with Nearly Everyone

As author Michael H. Hart discovered when he was doing research for his book, *The 100: A Ranking of the Most Influential Persons in History,* almost every significant discovery in the history of the world was masterminded by people living in cities. Many of the innovative individuals he wrote about journeyed from city to city—each time picking up fresh stimulation to develop more powerful ideas.

"Cities remain the cradle of civilization's creativity and ambition," notes Eugene Linden of *Time* magazine. "The catalytic mixing of people that fuels urban conflict also spurs the initiative, innovation, and collaboration that move civilization forward."

**Prediction:** Now that electronic methods for intensive communication—fax machines, computer modems, and cellular system relaying of telephone conversations through outer space—are becoming so widely available, cities will lose their historical edge in fostering creative innovation. Even people living in rural hamlets will have rich connections with distant experts who will help stimulate their creativity.

## Installing Your Favorite Telephone on Your Wrist

Advertisements circulated by AT&T recently announced that two-way Dick Tracey-style wrist telephones are on the way! When these compact and convenient little devices finally materialize (Fig. 12.2), can similar units linked to orbiting satellites be far behind?

**Prediction:** Before the last copy of this book disappears into oblivion, you will have a space-age telephone transceiver strapped to your wrist.

## Designing and Flying Tomorrow's Highly Efficient Booster Rockets

The extremely high cost of transporting payloads into orbit is the principal barrier to the orderly and inexpensive exploitation of the space frontier. Orbiting a pound of payload today costs $2000 or $3000, sometimes even more.

Yet my friend and colleague Ed Keith at Microcasm, Inc. has argued convincingly, in a number of different technical papers, that we should theoretically be able to orbit fairly large payloads for as little as $23 per pound. Ed Keith's proposed "asparagus stalk booster" (Fig. 12.3) consists of seven identical cylindrical stages strapped side by side to form a chubby little booster rocket. All seven of the cylindrical stages are powered by pressure-fed engines burning inexpensive kerosene and liquid oxygen; these engines are put together in large quantities on standardized mass-production assembly lines. Three times during each flight, pairs of rockets will be jettisoned until the seventh one drives its payload into space.

**Prediction:** The transportation of payloads into space, which today costs at least $2000 to $3000 per pound, will, within 7 years, drop to only about $500 per pound. Seven years after that, the price will be only $200 per pound.

## The Practical Value of Multifunctional Constellations

Today's satellite constellations are designed and installed one at a time in piecemeal fashion. Individually, the constellations usually operate rather well,

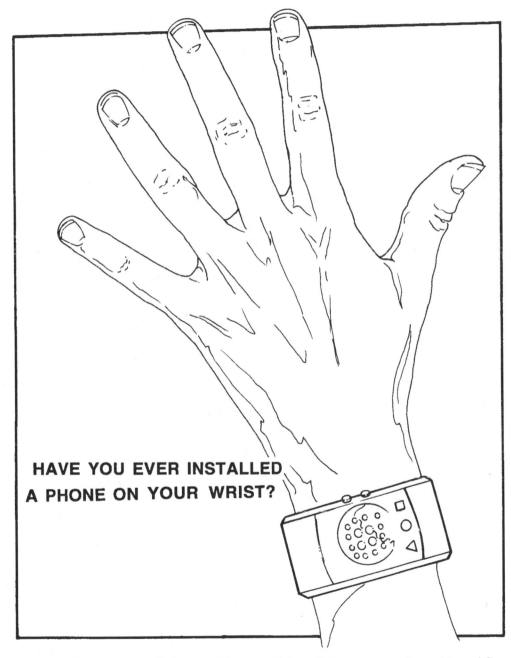

HAVE YOU EVER INSTALLED
A PHONE ON YOUR WRIST?

**Figure 12.2**   Two-way wrist telephones will become widely available as we move forward toward Century 21, and as electronic microminiaturization proceeds apace, wrist telephones for spaceborne voice messaging systems will inevitably follow. Larger video devices will also connect future users with various types of satellites—weather satellites, communication satellites, earth resources satellites, and satellites with ready access to computer-based archival data.

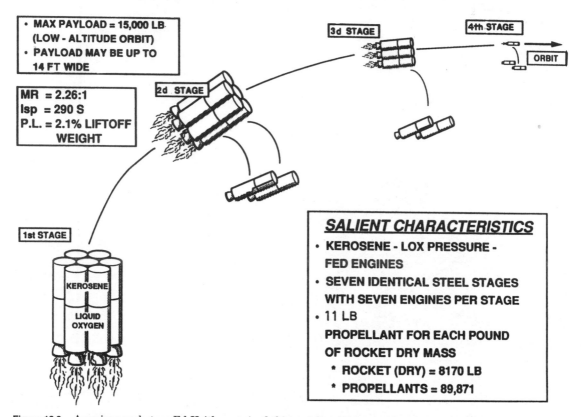

**Figure 12.3**   American rocketeer Ed Keith conceived this novel multistage rocket to serve as tomorrow's low-cost space-transportation system, which he believes will be able to deliver 15,000-lb cargoes into earth orbit for $22.78 per pound. Each of the seven parallel stages is fueled with inexpensive kerosene and liquid oxygen pressure fed into the combustion chamber. The multistage rocket drops two of the seven cylinders at each of the three staging points. The seventh cylinder then continues to burn until it drives its payload into the final destination orbit.

but their services are usually not well coordinated with one another. Users on the ground must employ special-purpose devices to gain access to each different type of constellation.

An alternate approach is depicted in Fig. 12.4. It involves careful coordination between constellations of satellites at three different orbital altitudes. These various satellites are planned and constructed with design commonality in mind so they can be a help to one another. In this particular case, the three subconstellations are positioned in low-altitude orbits, 12-hour orbits, and 24-hour (geosynchronous) orbits.

Reduced costs and interoperability are assured by designing the three constellations all at the same time. Crosslinking among the various satellites keeps all of them constantly coordinated with one another. Hand-held video-style devices owned and operated by nearly everyone will allow busy people everywhere access to all three constellations to obtain weather predictions,

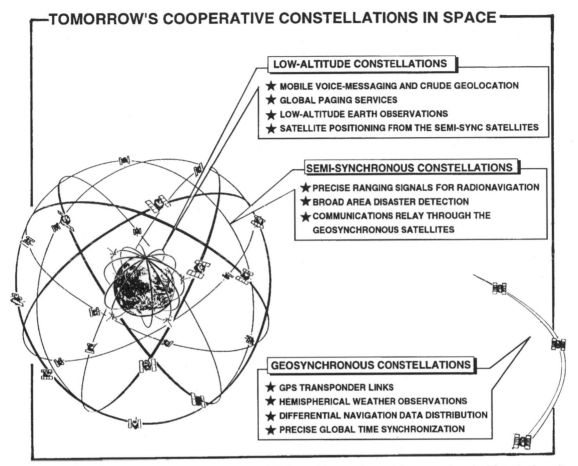

## TOMORROW'S COOPERATIVE CONSTELLATIONS IN SPACE

**LOW-ALTITUDE CONSTELLATIONS**
★ MOBILE VOICE-MESSAGING AND CRUDE GEOLOCATION
★ GLOBAL PAGING SERVICES
★ LOW-ALTITUDE EARTH OBSERVATIONS
★ SATELLITE POSITIONING FROM THE SEMI-SYNC SATELLITES

**SEMI-SYNCHRONOUS CONSTELLATIONS**
★ PRECISE RANGING SIGNALS FOR RADIONAVIGATION
★ BROAD AREA DISASTER DETECTION
★ COMMUNICATIONS RELAY THROUGH THE
  GEOSYNCHRONOUS SATELLITES

**GEOSYNCHRONOUS CONSTELLATIONS**
★ GPS TRANSPONDER LINKS
★ HEMISPHERICAL WEATHER OBSERVATIONS
★ DIFFERENTIAL NAVIGATION DATA DISTRIBUTION
★ PRECISE GLOBAL TIME SYNCHRONIZATION

**Figure 12.4** Future mobile communication systems will give ordinary private citizens, equipped with a single unit, instantaneous access to low-altitude, semisynchronous, and geosynchronous satellite constellations. The low-altitude satellites will furnish crystal-clear mobile communications, global paging, and crude geolocation services. The medium-altitude constellations will provide navigation solutions and broad-area disaster detection. Digital message relay, hemispherical weather observations, and precise time synchronization will be obtained in a similar manner from the geosynchronous satellites.

earth resources data, navigation positioning and routing, communication links, archival computer data, and the like.

Like the Ultimate Personal Computer, the video devices of the future to be used for accessing satellites can be operated in several different modes—with shared display screens, computer processing power, electronic storage devices, and common buttons that relabel themselves automatically whenever their user switches modes.

**Prediction:**   Tomorrow's highly capable constellations of satellites will be designed in groups with design commonality to provide an amazing variety of services through a multipurpose access unit owned by nearly every person on planet Earth.

## Finding Creative New Ways to Cope with Space Debris Fragments

Military technicians currently keep track of 7000 pieces of space debris as big as a soccer ball or bigger. Most, but not all, space debris fragments orbit within a few hundred miles of the surface of the earth (Fig. 12.5). If any of these objects ever happens to collide with a functioning satellite, the encounter could result in an explosive energy exchange.

Treaties to limit the amount of space debris will undoubtedly be signed as the space around the earth becomes increasingly crowded with space debris fragments intermingled with valuable satellites. Deorbiting the space debris using gentle radiation pressure induced by ground-based lasers pointing straight up toward the zenith will soon become a distinct possibility. The gentle upward pressure on the space debris gradually, over time, pushes it into a slightly elliptical orbit with its perigee dipping down into the thicker layers of the earth's atmosphere. Drag then nudges it into a reentry trajectory.

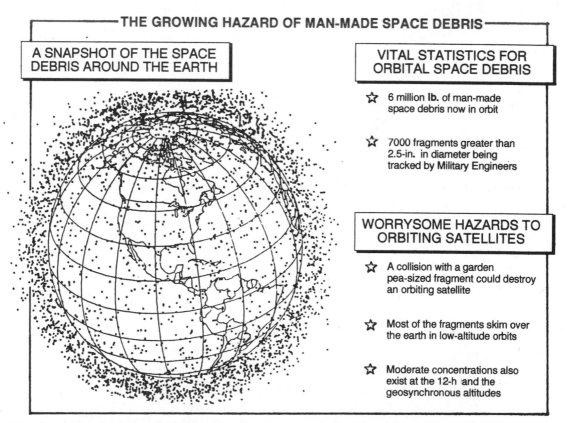

**THE GROWING HAZARD OF MAN-MADE SPACE DEBRIS**

### A SNAPSHOT OF THE SPACE DEBRIS AROUND THE EARTH

### VITAL STATISTICS FOR ORBITAL SPACE DEBRIS

☆ 6 million **lb.** of man-made space debris now in orbit

☆ 7000 fragments greater than 2.5-in. in diameter being tracked by Military Engineers

### WORRYSOME HAZARDS TO ORBITING SATELLITES

☆ A collision with a garden pea-sized fragment could destroy an orbiting satellite

☆ Most of the fragments skim over the earth in low-altitude orbits

☆ Moderate concentrations also exist at the 12-h and the geosynchronous altitudes

**Figure 12.5**  As spaceborne services become more and more popular, useless chunks of space debris will become increasingly hazardous to functioning satellites. Thus, in the next century, international control of space debris fragments will become a constant and costly preoccupation among the countries wanting to do business along the space frontier. The space debris fragments that clutter the low-altitude flight regime may even be deorbited by relatively gentle radiation pressure created by ground-based laser stations aiming their laser beams straight up toward the sky.

## HAZARDOUS ENCOUNTERS WITH ORBITAL SPACE DEBRIS

On July 2, 1982, during the final day of their mission, astronauts Ken Mattingly and Henry Hartsfield, riding the space shuttle *Columbia,* flew uncomfortably close to a spent Russian Intercosmos rocket high above the northwestern coast of Australia. By coincidence, that same region of space had experienced an earlier encounter with orbiting space debris when America's *Skylab* crashed in the outback in 1979. Astronauts Mattingly and Hartsfield were warned in advance, but they could not catch a glimpse of the big Intercosmos rocket as it whizzed by their spacecraft at 7000 mi/h.

Six months later, Russia's *Cosmos 1402* abruptly slammed into the earth. Like its sister ship *Cosmos 954,* it was a spy satellite powered by a nuclear reactor fueled with radioactive uranium. But, unlike its sister ship, *Cosmos 954* crashed to earth on the sovereign territory of an innocent nation. In 1978, when *Cosmos 954* fell in northern Canada, the Canadian government spent $6 million cleaning up the mess. Later, with some resistance, the Soviet Union reimbursed Canada for half that amount.

Military engineers track approximately 7000 objects in space as big as a soccer ball or bigger. A few hundred of them are functioning satellites. The rest are a varied lot: spent rockets, protective shrouds, clamps, fasteners, jagged fragments from space vehicle explosions, even an astronaut's silver glove. In addition to the 7000 objects of trackable size, tens of thousands of smaller ones are presently swarming around our planet.

These orbiting fragments are hazardous, but not to the people living on the ground below. On the average, human beings occupy the surface of the earth, only 17 tiny bodies per square mile. The *Skylab* was among the largest reentry bodies ever to plunge through the atmosphere, but scientific calculations indicated that the probability of any specific individual being hit by *Skylab* debris was only about 1 in 200 billion.

Actually, no calculations at all are needed to demonstrate that the probability of being bashed by orbital space debris is extremely small. More than 1500 large, hypervelocity meteorites are known to have plunged through the atmosphere and hit the earth—roughly 8 per year for the past 200 years. Many of them shattered into smaller fragments on reentry, but not one single human being's death certificate reads "death by meteorite." And yet, if we go back far enough into the dim shadows of history, we may find at least one reliable reference to human injury and death caused by falling meteorites. It is buried in the bible's book of Joshua, in a passage describing how terrified soldiers fleeing from battle were killed by "stones falling from heaven."

**Prediction:**    International treaties will help control hazardous space debris by the dawning of the 21st century. Spent stages and other useless fragments of space debris will be deorbited by radiation pressure induced by ground-based laser beams.

## Living, Loving, and Working in the Next Century, and Beyond

Those individuals fortunate enough to live in the 21st century need never lose touch with distant friends and loved ones because computers, coupled with fiber optic communication links and electromagnetic transmissions directed toward orbiting satellites, will give future earthlings unprecedented capabilities to communicate with anyone, anywhere they may choose to reach.

**Prediction:**    Tomorrow's innovative world will be equipped with marvelous new spaceborne mobile communication systems that will make daily life even richer and more exciting than the lives we now lead.

# Useful Magazines and Periodicals

*While theoretically and technically television may be feasible, commercially and financially I consider it an impossibility, a development of which we need waste little time dreaming.*

LEE DEFOREST
*Inventor of the vacuum tube amplifier, 1926*

Well-written magazines and periodicals can help you gain timely and accurate information on emerging technologies and promising employment opportunities in the design and operation of mobile communication satellites and their related ground equipment. *Space News,* for instance, and *Aviation Week* both publish a variety of articles on communication satellites and mobile communication systems. The industry trade journal, *Satellite Communications,* is a particularly interesting source of information on emerging technological trends and international financial arrangements. These, and the various other periodicals in the following list, will give you ample amounts of information from a variety of perspectives.

*Aerospace America*
370 L'Enfant Promenade SW
Washington, D.C. 20024
Phone: (202) 646-7400

*Air and Space*
Smithsonian Institution
900 Jefferson Drive
Washington, D.C. 20560
Phone: (202) 287-3733

*Aviation Week & Space Technology*
McGraw-Hill, Inc.
1221 Avenue of the Americas
New York, N.Y. 10020
Phone: (212) 512-2000

*Challenge*
Ball Aerospace Systems Group
Attn: Marketing Communications
P.O. Box 1062
Boulder, Colo. 80306-9818
Phone: (303) 939-4000

*GPS World*
859 Willamette Street
P.O. Box 10955
Eugene, Oreg. 97440-2460
Phone: (202) 783-4121

*Navigation: Journal of The Institute of Navigation*
815 15th Street NW, #832
Washington, D.C. 20005
Phone: (202) 783-4121

*The Journal of Navigation*
The Royal Institute of Navigation
1 Kensington Gore
London, SW7 2AT
England
Phone: 01-589-5021

*Scientific American*
415 Madison Avenue
New York, N.Y. 10017
Phone: (212) 754-0550

*Space News*
Army Times Publishing Co.
6883 Commercial Drive
Springfield, Va. 22159-0500
Phone: (703) 750-7400

*Defense News*
Army Times Publishing Co.
6883 Commercial Drive
Springfield, Va. 22159-0500
Phone: (703) 750-7400

*IEEE Communications*
Institute of Electrical and Electronics Engineers, Inc.
345 E. 47th Street
New York, N.Y. 10017-2394
Phone: (212) 705-7018

*AD ASTRA*
National Space Society
922 Pennsylvania Ave. SE
Washington, D.C. 20003-2140

*Microwave Journal*
Horizon House Publications, Inc.
685 Canton Street
Norwood, Mass. 02062
Phone: (617) 356-4595

*Spaceflight: The International Magazine of Space and Astronautics*
36 Washington Street
Glasgow, G3 8AZ
Scotland
United Kingdom

# B

# Professional Societies

*Railroad carriages are pulled at the enormous speed of 15 m.p.h. by engines which, in addition to endangering life and limb of passengers, roar and snort their way through the countryside, setting fires to the crops, scaring the livestock, and frightening women and children. The Almighty certainly never intended that people should travel at such breakneck speed.*

MARTIN VAN BUREN
*Eighth President of the United States, 1858*

Joining one or more of the world's many communications-related professional societies can be a stimulating and enjoyable way to gather additional information on spaceborne mobile communication systems. Your active participation in such an organization can also help you make important business and professional contacts. The organizations in this list welcome members with a variety of backgrounds, talents, and interests.

If you write or call, most of them will send free brochures describing their aims, activities, meetings, and publications. Use the materials that come back as a personal guide while you search for the perfect organization to join. The American Institute of Aeronautics and Astronautics, the Institute of Navigation, and the Royal Aeronautical Society have entirely different platforms (some of which are only peripherally related to mobile communication), but they are all stimulating and interesting organizations to join.

American Astronautical Society (AAS)
c/o University of Colorado
Campus Box 423
Boulder, Colo. 80309-0423
Phone: (703) 866-0020

American Institute of Aeronautics and
 Astronautics (AIAA)
370 L'Enfant Promenade SW
Washington, D.C. 20024
Phone: (202) 646-7400

British Interplanetary Society
27129 S. Lambeth Road
London, SW8 1SZ
England
Phone: 071-735-3160
Fax: 071-820-1504

Deutsche Gesellshaft fur Ortung
 un Navigation e.v.
Penpelforter Strasse 47
D-4000 Dusseldorf
Federal Republic of Germany
Phone: 49-211-369909
Fax: 49-211-351645

International Astronautics Federation
250 rue Saint Jacques
F-75005 Paris
France

Institute of Astrophysics
University of Liège
5 Avenue D / Cointe
B-4200 Cointe-Ougree
Liège, Belgium
Phone: 32-41-529980
Fax: 32-41-527474

Institute of Electrical and Electronics Engineers
 (IEEE)
445 Hoes Lane
Piscataway, N.J. 08855-1331
Phone: (908) 981-0060

Institute of Navigation (ION)
1026 16th Street NW, Suite 104
Washington, D.C. 20036
Phone: (202) 783-4121

Institute of Space
Instituto Italiano Di Dritto
Spaziale 251 via Giulia
Rome, Italy

Planetary Society
65 N. Catalina Avenue
Pasadena, Calif. 91106
Phone: (818) 793-5100

Royal Aeronautical Society
4 Hamilton Place
London, W1V 0BQ
England
Phone: 071-499-3515
Fax: 071-499-6230

Royal Institute of Navigation
1 Kensington Gore
London, SW7 2AT
England

# Bibliography

*The popular mind often pictures gigantic flying machines speeding across the Atlantic carrying innumerable passengers in a way analogous to our modern steam ships...it seems safe to say that such ideas are wholly visionary and even if the machine could get across with one or two passengers, the expense would be prohibitive to any but the capitalist who could use his own yacht.*

DR. WILLIAM H. PICKERING
*American astronomer, Massachussetts Institute of Technology*

## Chapter 1: Expanding Markets for Cellular Telephones

Bond, David F., "Mobile, Digital Audio Service Allocations Key to Upcoming World Radio Conference," *Aviation Week & Space Technology,* October 7, 1991, pp. 55–56.

Constable, George, *Communications,* Time-Life Books, Alexandria, Va., 1986.

"Cordless Communications," *The Economist,* October 9, 1993.

Deutsch, Sid, "Electromagnetic Field Cancer Scares," *Skeptical Inquirer,* Vol. 18, Winter 1993, pp. 152–156.

Deutschman, Alan, "The Next Big Info Tech Battle," *Fortune,* November 29, 1993, pp. 39–50.

Dixon, Robert C., *Spread Spectrum Systems,* John Wiley & Sons, New York, 1984.

Elmer-Dewitt, Philip, "First Nation in Cyberspace," *Time,* December 6, 1993, pp. 62–64.

Groves, Martha, "Pactel to Launch New Cellular Phone System," *Los Angeles Times,* January 13, 1994, p. D3.

Heldman, Robert K., *Future Telecommunications: Information Applications, Services, & Infrastructure,* McGraw-Hill, New York, 1993.

Helm, Leslie, "Fleet Thinking Is Helping Tiny Nextel Make Big Waves," *Los Angeles Times,* December 5, 1993, pp. D1, 3.

Kupfer, Andrew, "Look, Ma! No Wires!", *Fortune,* December 13, 1993, pp. 147–152.

Kupfer, Andrew, "Phones That Will Work Anywhere," *Fortune,* August 24, 1992, pp. 100–112.

Kupfer, Andrew, "The Go-Anywhere Phone Is At Hand," *Fortune,* November 5, 1990, pp. 142–145.

Lazzareschi, Carla, "Is Phone's Next Step a New Mother Lode?" *Los Angeles Times,* September 7, 1993, pp. D1, 12.

Lee, William C. Y., *Mobile Cellular Telecommunications Systems,* McGraw-Hill, New York, 1989.

Logsdon, Tom, *Breaking Through: Creative Problem Solving Using Six Successful Strategies,* Addison-Wesley, Reading, Mass., 1993.

Macario, Raymond C. V., *Cellular Radio—Principles and Design,* McGraw-Hill, New York, 1993.

McCarroll, Thomas, "The Humongous Hookup," *Time,* August 30, 1993, pp. 33–34.

Mitchell, Peter W., "Digitizing Sound," *Technology Illustrated,* October/November 1981, pp. 26–29.

Porter, Glenn, "When Telephony Was Young," *Science,* Vol. 259, January 29, 1993, pp. 699–701.

Ridley, Randy, "EMF: Truth or Scare?" *Satellite Communications,* May 1993, pp. 44–45.

Schrage, Michael, "This Road Will Be a Tough One to Pave," *Los Angeles Times,* January 13, 1994, pp. D1, 5.

Seitz, Patrick, "Cancer Scare No Threat to Plans," *Space News,* November 23–29, 1992, p. 21.

Shell, Ellen Ruppel, "Bach in Bits," *Technology Illustrated,* October/November 1981, pp. 21–26.

Solomon, Jolie, and Jennifer Foote, "Reach Out and Buy Someone," *Newsweek,* June 14, 1993, p. 49.

Sweeney, Dan, "Megaplayers and Consortia Are Aligning Themselves to Compete for Satellite-Based Mobile Communications Business," *Cellular Business,* July 1993, pp. 1–5.

Therrien, Lois, "A Double Standard for Cellular Phones," *Business Week,* April 27, 1992, pp. 104–105.

"Tokyo Answers the Call," *Time,* July 10, 1989, p. 46.

Wertz, James R., and Wiley J. Larson (Editors), *Space Mission Analysis and Design,* Kluwer Academic Publishers, Dordrecht, The Netherlands, 1991.

# Chapter 2: Emerging Trends in Communication Satellites

Aghvami, A. H., "Digital Modulation Techniques for Mobile and Personal Communication Systems," *Electronics Communication,* June 1993, pp. 125–132.

Alper, Joel, and Joseph N. Pelton, *The Intelsat Global Satellite System,* American Institute of Aeronautics and Astronautics, New York, 1984.

Bond, David F., "Mobile, Digital Audio Service Allocations Key to Upcoming World Radio Conference," *Aviation Week & Space Technology,* October 7, 1991.

"Conceptual Communication Services Satellite," *The Foundation Report,* June 1, 1978, p. 35.

Constable, George, *Communications,* Time-Life Books, Alexandria, Va., 1986.

Curtis, Anthony R., *Space Almanac,* Gulf Publishing, Houston, Tex., 1992.

Damon, Thomas, *Introduction to Space: The Science of Spaceflight,* Orbit Book Company, Malabar, Fla., 1990.

Edelson, Burton I., et al., *NASA/NSF Panel Report on Satellite Communications Systems and Technology: Volume I. Analytical Chapters,* International Technology Research Institute, Baltimore, Md., 1993.

Goetz, I., "Ensuring Mobile Communication Transmission Quality in the Year 2000," *Electronics Communication,* June 1993, pp. 141–146.

Helm, Leslie, "The World's Networks," *Los Angeles Times,* October 24, 1993, p. D1+.

Logsdon, Tom, *Computers Today and Tomorrow: The Microcomputer Explosion,* Computer Science Press, Rockville, Md., 1985.

Logsdon, Tom, "New Space Initiatives," *IR&D Report SSD 104 77,* Rockwell International, September 19, 1978, pp. 1–32.

Logsdon, Tom, "Orbiting Switchboards," *Technology Illustrated,* October/November 1981, pp. 55–62.

Logsdon, Tom, *Space Inc., Your Guide to Investing in Space Exploration,* Crown Publishers, New York, 1988.

Logsdon, Tom, *The Navstar Global Positioning System,* Van Nostrand Reinhold, New York, 1992.

Logsdon, Tom, *The Rush Toward the Stars,* Franklin, Englewood, N.J., 1970.

Manuta, Lou, "Now Is the Winter of Our Dish Content," *Satellite Communications,* February 1993, pp. 21–23.

Pattan, Bruno, *Satellite Systems: Principles and Technologies,* Van Nostrand Reinhold, New York, 1993.

Pelton, Joseph N., *The "How To" of Satellite Communications,* Design Publishers, Sonoma, Calif., 1991.

Seitz, Patrick, "Going, Going, Gone: Selling Cars Via Satellite," *Space News,* November 23–29, 1992.

Seitz, Patrick, "PanAmSat Nears Fullment of Vision," *Space News,* September 27–October 3, 1993.

Seitz, Patrick, "Satellites Deliver Educational TV," *Space News,* August 2–8, 1993.

"Space Industrialization," *Executive Summary Report: Rockwell International,* April 14, 1978, pp. 1–39.

Webb, W.T., "Sizing Up the Microcell for Mobile Radio Communications," *Electronics Communication,* June 1993, pp. 133–140.

Wertz, James R., and Wiley J. Larson (Editors), *Space Mission Analysis and Design,* Kluwer Academic Publishers, Dordrecht, The Netherlands, 1991.

Wu, William W., *Elements of Digital Satellite Communication,* Computer Science Press, Rockville, Md., 1985.

Yenne, Bill, *The Encyclopedia of US Spacecraft,* Exeter Books, New York, 1985.

# Chapter 3: Mobile Communications on the High Seas

"A Vision of Tomorrow's World," *Ocean Voice,* January 1993, pp. 23–25.

"Catching Data," *Ocean Voice,* July 1992, pp. 35–38.

de Selding, Peter B., "Inmarsat Outlines Trio of Plans for Project 21 Satellite System," *Space News,* March 29–April 4, 1993, p. 14.

de Selding, Peter B., "Inmarsat, Salyut Ready to Sign Contract," *Space News,* April 19–25, 1993, pp. 3, 29.

Edelson, Burton I., and Joseph N. Pelton, "Satellite Communication Systems and Technology," *Executive Summary: International Technology Research Institute,* Report, July 1993, pp. 1–8.

Holtzman, Robert, "Special Delivery," *Ocean Voice,* January 1993, pp. 36–37.

Logsdon, Tom, *Space Inc.: Your Guide to Investing in Space Exploration,* Crown Publishers, New York, 1988.

"Mobility in Miniature," *Ocean Voice,* October 1992, pp. 13–17.

Taillade, Michel, "Focus on Fishing," *Mobile Monitoring Newsletter,* January 1993, pp. 2–12.

"The Argos System: A General Presentation," *Argos,* May 1993, pp. 1–7.

"The Global Phone 2000t," *Satellite Communications,* October 1992, pp. 42–43.

# Chapter 4: Spaceborne Land-Mobile Communication Systems

Dewar, Janet, and Martha Cooley, "Chips, Bytes and Birds," *Satellite Communications,* August 1993, pp. 26ff.

Holzmann, Gerard J., and Bjorn Pehrson, "The First Data Networks," *Scientific American,* January 1994, pp. 124–129.

Jacobs, Irwin M., Allen Salmasi, and Thomas J. Bernard, "OmniTRACS Technical Overview," Qualcomm OmniTRACS booklet, San Diego, Calif., February 1991, pp. 1–17.

Lawton, George, "Paging: The Mouse that Roared," *Satellite Communications,* September 1992, pp. 29–32.

Marek, Sue, "Paying for Portability," *Satellite Communications,* July 1993, pp. 34–39.

McCarroll, Thomas, "Betting on the Sky," *Time,* November 22, 1993, p. 57.

Nordwall, Bruce D., "Mobile Communications to Capture Consumer Market," *Aviation Week & Space Technology,* May 31, 1993, pp. 41–42.

*OmniTRACS Two-Way Satellite-Based Mobile Communications*, Qualcomm OmniTRACS booklet, San Diego, Calif., 1994.

*QDSS: The Qualcomm Decision Support System,* Qualcomm OmniTRACS booklet, San Diego, Calif., 1994.

"Satellite Tracking of Hazardous Materials," Argos booklet, Paris, 1994.

Seitz, Patrick, "New RDSS Firm Emerges," *Space News,* March 29–April 4, 1993, p. 18.

*VIS: OmniTRACS Vehicle Information Systems,* Qualcomm OmniTRACS booklet, San Diego, Calif., 1994.

# Chapter 5: Satellite Messaging for Commercial Jets

Asker, James R., "Inmarsat Braces for Multiple Rivals," *Aviation Week & Space Technology,* October 11, 1993, pp. 44–45.

Brandli, Hank, "APT on the Wing," *Satellite Communications,* November 1993, pp. 2A–3A.

Covault, Craig, "European Airlines Adopting Satellite Technology," *Aviation Week & Space Technology,* November 23, 1992, pp. 44–45.

"In-Flight Phone to Supply Entertainment Radio," *Aviation Week & Space Technology,* March 1, 1993, p. 33.

Klass, Philip J., "Inmarsat Orbit to Impact Aviation," *Aviation Week & Space Technology*, March 29, 1993, pp. 57, 59.

Klass, Philip J., "Satellites to Become Primary Link Between Airline Transports, Ground," *Aviation Week & Space Technology,* April 1, 1991, pp. 40–43.

Klass, Philip J., "Satellites to Revolutionize Trans-Oceanic Operations," *Aviation Week & Space Technology,* November 25, 1991, pp. 93–96.

Klass, Philip J., "Wider Use of Satellite Communications in Aircraft to Benefit Airlines, Passengers," *Aviation Week & Space Technology,* August 3, 1992, pp. 70–71.

Nordwall, Bruce D., "Digital Data Links Key to ATC Modernization," *Aviation Week & Space Technology,* January 10, 1994, pp. 53–55.

"Orbiting Data Link Aids Ocean-Crossing Airlines," *Signal,* September 1993, pp. 37–39.

Ott, James, "New Technology Key to Airline Cost Cuts," *Aviation Week & Space Technology,* November 23, 1992, pp. 42–43.

Pattan, Bruno, *Satellite Systems: Principles and Technologies,* Van Nostrand Reinhold, New York, 1993.

Reller, Jane, "Dial Direct from the Friendly Skies," *Inside Track Newsletter,* 1993.

Riccitiello, Robina, "Airlines' Interest Grows in Satellite Phone Service," *Space News,* November 16–22, 1992, pp. 15, 23.

Riccitiello, Robina, "Airlines' Use of Satellites May Triple," *Space News,* May 3–9, 1993, p. 16.

Riccitiello, Robina, "Satellites Allow Aircraft Flight Plan Revisions Mid-Flight," *Space News,* May 17, 1993, p. 23.

Riccitiello, Robina, "Satellites Modernize Air Traffic Control, Communications," *Space News* November 16–22, 1992, pp. 14–15.

Robinson, Clarence A., Jr., "Rapid-Fire Digital Link Offers Striking Results," *Signal,* September 1993, pp. 15–18.

Shifrin, Carole A., "European Airlines Evaluate Land-Based Phone Service," *Aviation Week & Space Technology,* October 11, 1993, pp. 41–43.

Stahr, Scott, "Flying the (User) Friendly Skies," *Satellite Communications,* November 1993, pp. 4A–5A.

"United to Outfit New, Long-Range Transports With Rockwell/Collins Communications Systems," *Aviation Week & Space Technology,* April 20, 1992, p. 39.

## Chapter 6: Selecting the Proper Constellation Architecture

Baker, Robert M., Jr., and Maud W. Makemson, *An Introduction to Astrodynamics,* Academic Press, New York, 1960.

Battin, Richard H., *An Introduction to the Mathematics and Methods of Astrodynamics,* American Institute of Aeronautics and Astronautics, New York, 1987.

Bink, Barbara H., "The Impact of Mobile and Wireless Communications on Business," *Mobile Product News,* 1994, pp. 1–15.

Bryan, C.D.B., *The National Air and Space Museum,* Harry N. Abrams, New York, 1984.

Farmer, Dean, "Using Today's Strategic Defense Initiative (SDI) Technologies to Accomplish Tomorrow's Low Cost Space Missions," *World Space Congress,* August 31, 1992, pp. 1–10.

Frieden, Rob, "Much Ado About Bandwidth," *Satellite Communications,* February 1993, pp. 24–28.

Indrikis, Janis, and Robert Cleave, "Space Eggs," 5th Annual Utah State University Conference on Small Satellites, *Proceedings of the 5th Annual AIAA/USU Conference at Utah State University,* August 28, 1991, pp. 1–7.

Keith, Edward L., "Low-Cost Space Transportation: Economic Impact and Applications," 10th Annual Space Development Conference, May 1991, pp. 1–23.

Keith, Edward L., "Low-Cost Space Transportation: The Search for the Lowest Cost," American Astronautical Society/American Institute of Aeronautics and Astronautics, technical paper, February 13, 1991, pp. 1–20.

Logsdon, Tom, *Computers Today and Tomorrow: The Microcomputer Explosion,* Computer Science Press, Rockville, Md., 1985.

Pattan, Bruno, *Satellite Systems: Principles and Technologies,* Van Nostrand Reinhold, New York, 1993.

Rusch, Roger J., *Comparison of Personal Communications Satellite Systems,*" TRW Inc., Redondo Beach, Calif., 1993, pp. 1–10.

Wertz, James R., and Wiley J. Larson (Editors), *Space Mission Analysis and Design,* Kluwer Academic Publishers, Dordrecht, The Netherlands, 1991.

Yenne, Bill, *The Encyclopedia of US Spacecraft,* Exeter Books, New York, 1985.

# Chapter 7: Mobile Communication Satellites at Geosync

Alper, Joel, and Joseph N. Pelton, *The Intelsat Global Satellite System*, American Institute of Aeronautics and Astronautics, New York, 1984.

"America's Mobile Satellite," American Mobile Satellite booklet, 1993.

Baranowsky, Patrick W., II, "MSAT and Cellular Hybrid Networking," *American Mobile Satellite*, 1993, pp. 149–154.

Barrett, Randy, "All Systems Go for AMSC," *Washington Technology: The Business Newspaper of Technology*, March 25, 1993, pp. 1–2.

Connelly, Brian, "The Westinghouse Series 1000 Mobile Phone: Technology and Applications," *Westinghouse Electronics Systems*, 1993, pp. 375–379.

Hess, Elizabeth, "Project 21: Leo, Meo or Geo?" *Satellite Communications,* October 1993, pp. 42–46.

Johanson, Gary A., N. George Davies, and William R. H. Tisdale, "Implementation of a System to Provide Mobile Satellite Services in North America," *American Mobile Satellite*, 1993, pp. 279–284.

Johnson, Nicholas L., "The Soviet Year in Space: 1990," Teledyne Brown Engineering, 1991, pp. 1–49.

Logsdon, Tom, *Computers Today and Tomorrow: The Microcomputer Explosion*, Computer Science Press, Rockville, Md., 1985.

Logsdon, Tom, *The Rush Toward the Stars*, Franklin, Englewood, N.J., 1970.

Lunsford, J., R. Thorne, D. Gokhale, W. Garner, and G. Davies, "The AMSC/TMI Satellite Services System: Ground Segment Architecture," American Institute of Aeronautics and Astronautics, Dec. 1992, pp. 405–425.

Pattan, Bruno, *Satellite Systems: Principles and Technologies*, Van Nostrand Reinhold, New York, 1993.

Pelton, Joseph, "Will SmallSat Markets Be Large?" *Satellite Communications*, February 1993, pp. 30–42.

Reichhardt, Tony, "Little Launches: Small Spacecraft Are Becoming the Wave of the Present," *Air & Space*, June–July 1993, pp. 31–35.

"Skycell Fleet Management," American Mobile Satellite booklet, April 1993.

"Special Report: Mobile Satellite Service Will Launch Wireless Communications Into the Global Frontier," *American Mobile Satellite*, June 7, 1993, pp. 9–12.

"The Executive Report on MSS Voice and Data Systems," *Mobile Satellite News,* June 23, 1993, pp. 1–6.

Tucci, Liz, "Frequency Distribution Compromise Proposed," *Space News,* January 24–30, 1994, pp. 1–2.

Whalen, David J., and Gary Churan, "The American Mobile Satellite Corporation Space Segment," International Mobile Satellite Conference, June 16–18, 1993, pp. 394–403.

White, Lawrence, Anil Agarwal, Brian Skerry, and Bill Tisdale, "North American Mobile Satellite System Signaling Architecture," American Institute of Aeronautics and Astronautics, Dec. 1992, pp. 427–439.

# Chapter 8: Telegram Satellites Launched into Low-Altitude Orbits

Asker, James R., "Orbcomm Satellites to Use Unique Disk Shape," *Aviation Week & Space Technology,* September 27, 1993, pp. 49–51.

Asker, James R., "Upstart Satellite Companies Press for New Telecommunications World Order," *Aviation Week & Space Technology,* October 7, 1991, pp. 50–54.

Barresi, G., G. Rondinelli, C. Soddu, and D. L. Brown, "Leostar: A Small Spacecraft for LEO Communication Missions," *Italspazio,* 1993, pp. 1–12.

Childs, John, "OSC/Hercules Pegasus," *Aviation Week & Space Technology,* September 3, 1990, pp. S3–S10.

Dixon, Robert C., *Spread Spectrum Systems,* John Wiley & Sons, New York, 1984.

Dorfman, Steve, "Satellites: Key to Success: Relying Solely on Fiber-Optic Cable Is Too Costly," *Space News,* February 21, 1994, p. 16.

Ellis, Janius, "Entrepreneurs in Space," *Air and Space,* December 1986/January 1987, pp. 98–101.

Hartshorn, David, "Target Market Earth," *Satellite Communications,* October 1992, pp. 26–32.

Hatlelid, John E., and David E. Sterling, "A Survey of Small Spacecraft in Commercial Constellations," *Motorola Inc.,* 1991, pp. 2–7.

Indrikis, Janis, and Robert Cleave, "Space Eggs," 5th Annual Utah State University Conference on Small Satellites, *Proceedings of the 5th Annual AIAA/USU Conference at Utah State University,* August 28, 1991, pp. 1–7.

Kaveeshwar, Ashok, The STARSYS Data Messaging and Geo-Positioning System," STARSYS Global Positioning, Inc. booklet, Lanham, Mass., September 1993, pp. 1–10.

Klass, Philip J., "Low-Earth Orbit Communications Satellites Compete for Investors and U.S. Approval," *Aviation Week & Space Technology,* May 18, 1992, pp. 60–61.

Klass, Philip J., "WARC-92 Approves Satellites for Small Cellular Telephones," *Aviation Week & Space Technology,* March 9, 1992, p. 31.

Manuta, Lou, "It's A Boy! Little LEO Joins the Family," *Satellite Communications,* January 1994, p. 19.

Nordwall, Bruce D., "Mobile Communications to Capture Consumer Market," *Aviation Week & Space Technology,* May 31, 1993, pp. 41–42.

Pattan, Bruno, *Satellite Systems: Principles and Technologies,* Van Nostrand Reinhold, New York, 1993.

Pelton, Joseph N., *The "How To" of Satellite Communications,* Design Publishers, Sonoma, Calif., 1991.

Renshaw, Alan B., "The STARSYS Global Positioning and Messaging System," STARSYS Global Positioning, Inc., Lanham, Mass., 1992, pp. 1–10.

Seitz, Patrick, "Little LEO Firms Trying to Exhibit Individuality," *Space News,* August 9–15, 1993, p. 10.

Seitz, Patrick, "Little LEO's Win Unanimous Vote of Commission," *Space News,* October 25–31, 1993, pp. 4, 21.

Seitz, Patrick, "Orbcomm Design Unveiled: Sights Set High for System's Earning Potential," *Space News,* August 30–September 5, 1993, p. 1.

STARSYS, STARSYS Global Positioning, Inc. booklet, Lanham, Mass., 1994.

Tucci, Liz, "Report Says NASA Should Drop Spacehab," *Space News,* October 11–17, 1993, pp. 1–20.

## Chapter 9: Big Constellations in Low-Altitude Orbits

"Always on Call: Motorola Hopes to Connect the Globe With Cellular Phones," *Time,* July 9, 1990, p. 51.

Asker, James R., "Inmarsat Braces for Multiple Rivals," *Aviation Week & Space Technology,* October 11, 1993, pp. 44–45.

Barrett, Randy, "Calling Corp. to Use 840 Sats in LEO," *Washington Technology: The Business Newspaper of Technology,* July 15, 1993, p. 1, 5.

Bertiger, Bary R., and Peter A. Swan, "Iridium Is in the Works," *Aerospace America,* February 1991, pp. 40–42.

Chase, Scott, "Iridium Gets the Call," *Ad Astra,* January 1994, pp. 38–42.

de Selding, Peter B., and Liz Tucci, "Motorola Seeks to Block Comsat," *Space News,* July 19–21, 1993, pp. 3, 29.

de Selding, Peter B., "Globalstar Team Unveils Partners in 48-Satellite Spacecraft Telephone Plan," *Space News,* March 28, 1994, pp. 1+.

de Selding, Peter B., "Schwartz Lists Globalstar Backers," *Space News,* March 28, 1994, p. 20.

Dixon, Robert C., *Spread Spectrum Systems,* John Wiley & Sons, New York, 1984.

Frieden, Rob, "Get a Grip...Or Lose a Market," *Satellite Communications,* July 1993, pp. 40–41.

Gellene, Denise, "Private System Linking Globe by Satellite Planned," *Los Angeles Times*, March 21, 1994, pp. A1+.

Gipson, Melinda, "Calling Communications' Financial Details Sketchy," *Space News,* September 27–October 3, 1993, p. 20.

"Global Telephony Network of 840 LEO Satellites Planned," *Telecommunications Reports,* June 21, 1993, pp. 36–37.

Globalstar System: General Summary, Loral Qualcomm Satellite Services, Inc. booklet, New York, 1994, pp. 1–19.

Globalstar, Loral Qualcomm Satellite Services Inc. booklet, New York, 1994, pp. 1–12.

Hatfield, Dale N., "A Discussion of the Globalstar Satellite-Based Mobile Telecommunications System," Hatfield Associates: Executive Summary, booklet, New York, 1992, pp. 1–7.

Hendrickson, Rich and Kent Penwarden, Globalstar for the Military, Loral Qualcomm Satellite Services, Inc. booklet, New York, 1992, pp. 1–12.

Hill, Christian, and Ken Yamada, "Staying Power: Motorola Illustrates How an Aged Giant Can Remain Vibrant," *Wall Street Journal,* December 9, 1992, pp. 1–3.

Kantrowitz, Barbara, and Daniel McGinn, "Bill and Craig's Big Idea: Wiring the World with 840 Satellites," *Newsweek,* April 4, 1994, p. 44.

Kim, James, and Gary Strauss, "Stock Deal a 'Nightmare' for Rivals," *USA Today,* August 17, 1993, B1+.

Klass, Philip J., "Low Earth Orbit Communications Satellites Compete for Investors and U.S. Approval," *Aviation Week & Space Technology,* May 18, 1992, pp. 60–61.

Lenorovitz, Jeffrey M., "Energia Gains U.S. Backing for Signal Satellite System," *Aviation Week & Space Technology,* February 28, 1994, pp. 55–57.

Marcus, Daniel J., and Peter B. de Selding, "Khrunkhev to Invest $40 Million in Iridium," *Space News,* January 18–24, 1993, pp. 3, 21.

Marek, Sue, "Edward Tuck's Calling," *Satellite Communications,* August 1993, p. 10.

Morant, Andrew, "High-Powered Ideas," *Financial Times,* October 15, 1992, p. 13.

Navarra, Anthony, "Globalstar," Mobile Communications International, booklet, New York, 1993, pp. 46–48.

Peltz, James F., "How TRW Came Down to Earth," *Los Angeles Times,* February 21, 1994, pp. D1+.

Peltz, James F., "Network Idea Man," *Los Angeles Times,* March 23, 1994, pp. D1+.

"Phones Into Orbit," *The Economist,* March 28–April 3, 1992, p. 15.

Seitz, Patrick, "Motorola Reveals Equity Investors for Iridium Project," *Space News,* August 9–15, 1993, p. 11.

Seitz, Patrick, "New LEO Satellite Venture Proposes Ambitious Plans," *Space News,* June 28–July 11, 1993, p. 16.

Seitz, Patrick, "Scarce Spectrum Drives New Users to Ka-Band," *Space News,* March 7–13, 1994, p. 20.

Shiver, Jube, Jr., "Battle to Control the Skies," *Los Angeles Times,* March 22, 1994, pp. D1+.

Slutsker, Gary, "The Company That Likes to Obsolete Itself," *Forbes,* September 13, 1993, pp. 139–144.

Therrien, Lois, "Motorola's Iridium: A Long Slog to the Launchpad," *Business Week,* November 23, 1992, pp. 116–120.

Tucci, Liz, "Firm Plans 840-Mobile-Satellite Constellation," *Space News,* January 25–31, 1993, p. 1, 20.

Tucci, Liz, "Teledesic System Proposal Awaits FCC Approval," *Space News,* March 28, 1994, p. 2.

Tuck, Edward F., David P. Patterson, James R. Stuard, and Mark H. Lawrence, "The Calling Network: A Global Telephone Utility," Calling Communications booklet, Newport Beach, Calif., 1993, pp. 1–26.

Tuck, Edward F., "The First MegaLEO?" *Satellite Communications,* November 1993, pp. 20–22.

Werner, Debra Polsky, "Lockheed to Sell Generic Satellite Based on Iridium," *Space News,* January 24–30, 1994, p. 24.

## Chapter 10: Medium-Altitude Constellations

Broaiua, J. W., and D. Castiel, The Ellipso Mobile Satellite System, Ellipsat Corp. booklet, Washington, D.C., 1993.

Bryan, C. D. B., *The National Air and Space Museum,* Harry N. Abrams, New York, 1984.

de Selding, Peter B., "Inmarsat Weighs Two Alternative Proposals for Its Mobile Service," *Space News,* August 9–15, 1993, p. 8.

Draim, J., *Lightsat Constellation Designs,* American Institute of Aeronautics and Astronautics, 1992, pp. 1361–1369.

Draim, John E., and Thomas J. Kacena, "Populating the Abyss—Investigating More Efficient Orbits: Getting More Miles to the Gallon for Your (Space) Vehicle," Space Applications Corp. prepublication booklet, 1992, pp. 1–15.

Farmer, Dean, *Using Today's Strategic Defense Initiative (SDI) Technologies to Accomplish Tomorrow's Low Cost Space Missions,* World Space Congress, August 31, 1992, pp. 1–10.

Indrikis, Janis, and Robert Cleave, "Space Eggs," 5th Annual Utah State University Conference on Small Satellites, *Proceedings of the 5th Annual AIAA/USU Conference at Utah State University,* August 28, 1991, pp. 1–7.

McDonald, Keith, and Kim Nussbaum, The Feasibility of Using Small, Low-Cost Satellites (Econosats) as an Augmentation to GNSS and as an Eventual Fully Capable GNSS, Sat Tech Systems, Inc., Arlington, Va., 1993, pp. 1–8.

Pattan, Bruno, *Satellite Systems: Principles and Technologies,* Van Nostrand Reinhold, New York, 1993.

Rusch, J., Comparison of Personal Communications Satellite Systems, TRW, Inc. booklet, Redondo Beach, Calif., 1993, pp. 1–10.

Rusch, Roger J., *Odyssey, An Optimized Personal Communications Satellite System,* International Astronautical Federation, Redondo Beach, Calif., October 1993, pp. 1–11.

Rusch, Roger J., Peter Cress, Michael Horstein, Robert Huang, and Eric Wisell, *Odyssey, A Constellation for Personal Communications,* American Institute of Aeronautics and Astronautics, Washington, D.C., 1992, pp. 1–10.

Spitzer, Christopher J., Odyssey Personal Communications Satellite System, 15th Pacific Telecommunication Conference, January 17–20, 1993, pp. 1–13.

Yenne, Bill, *Encyclopedia of US Spacecraft,* Exeter Books, New York, 1985.

## Chapter 11: Summing Up

Augustine, Norman R., *Augustine's Laws,* American Institute of Aeronautics and Astronautics, New York, 1983.

Constable, George, *Communications,* Time-Life Books, Alexandria, Va., 1986.

David, Leonard, "Public Demonstrations Show Promise of Telepresence," *Space News,* November 29–December 5, 1993.

Kessler, D. J., and B. G. Cour-Palais, "Collision Frequency of Artificial Satellites: The Creation of a Debris Belt," *Journal of Geophysical Research,* Volume 83, 1979, pp. 2637–2646.

Kupfer, Andrew, "The Rise to Rewire America," *Fortune,* April 19, 1993, pp. 42–61.

Logsdon, Tom, *The Navstar Global Positioning System,* Van Nostrand Reinhold, New York, 1992.

Maurer, Richard, *Junk in Space,* Simon and Schuster, New York, 1989.

Pattan, Bruno, *Satellite Systems: Principles and Technologies,* Van Nostrand Reinhold, New York, 1993.

Pelton, Joseph N., *The "How To" of Satellite Communications,* Design Publishers, Sonoma, Calif., 1991.

Ridley, Randy, "Is Fiber Getting Out of Hand?" *Satellite Communications,* August 1993, pp. 22–24.

Rusch, Roger J., Comparison of Personal Communication Satellite Systems, TRW, Inc. booklet, Redondo Beach, Calif., 1993, pp. 1–10.

## Chapter 12: Crystal Ball Predictions for Century 21

"A Model for Low Earth Orbit?" *Space News,* March 28, 1994, p. 14.

Augustine, Norman R., *Augustine's Laws,* American Institute of Aeronautics and Astronautics, New York, 1983.

Chen, Edwin, "Fearing Collisions in Space: U.S. Tracks Orbiting Debris," *Los Angeles Times,* December 25, 1993, pp. A1+.

de Selding, Peter B., "For Europe, Exports in Jeopardy if U.S. Ventures Sew Up Market," *Space News,* March 28, 1994, pp. 1+.

Frieden, Rob, "Get a Grip...or Lose a Market," *Satellite Communications,* July 1993, pp. 40–41.

Gellene, Melinda, "Private System Linking Globe by Satellite Planned," *Los Angeles Times,* March 21, 1994, pp. A1+.

Hart, Michael H., *The 100: A Ranking of the Most Influential Persons in History,* Citadel Press, Secaucus, N.J., 1987.

Klass, Phillip J., "Low-Earth Orbit Communications Satellites Compete for Investors and Approval," *Aviation Week & Space Technology,* May 18, 1992, pp. 60–61.

Linden, Eugene, "Megacities," *Time,* January 11, 1993, pp. 28–38.

Logsdon, Tom, *Breaking Through: Creative Problem Solving Using Six Successful Strategies,* Addison-Wesley, Reading, Mass., 1993.

Logsdon, Tom, *Our Computerized Society with Basic Programming,* Anaheim Publishing Company, Fullerton, Calif., 1979.

Nown, Graham, *The World's Worst Predictions,* Arrow Books, London, England, 1985.

Seitz, Patrick, "Little Leo Firms Trying to Exhibit Individuality," *Space News,* August 9–15, 1993, pp. 4, 21.

Stern, Jill Abeshouse, and LaRene Tondro, "A Space-Based Information Highway," *Space News,* November 8–14, p.15.

*Welcome to the Untethered World,* booklet, Deloitte and Touche, Toronto, Canada, 1993.

Wertz, James R., and Wiley J. Larson (Editors), *Space Mission Analysis and Design,* Kluwer Academic Publishers, Dordrecht, The Netherlands, 1991.

Wu, William W., *Elements of Digital Satellite Communication,* Computer Science Press, Rockville, Md., 1985.

# Glossary

*There is no hope for the fanciful idea of reaching the moon, because of insurmountable barriers to escaping the earth's gravity.*

DR. F.R. MOULTON
*Astronomer, University of Chicago, 1932*

**ACCELERATION**   Any change in velocity.

**ACOUSTIC COUPLER**   An electronic device that fosters the efficient transmission of digital data over ordinary analog telephone circuits.

**AMPLIFIER**   Any device used in boosting the strength of an electronic signal.

**ANALOG COMPUTER (see also DIGITAL COMPUTER)**   A computer whose processing functions are based on measuring continuous variables rather than counting discrete entities.

**ANALOG MODULATION (see also DIGITAL MODULATION)**   The process of manipulating the frequency or the amplitude of an electromagnetic carrier wave by continuously variable amounts, as opposed to digital modulation in which any variations occur in discrete numerical steps.

**ANALOG-TO-DIGITAL CONVERSION (see also DIGITAL-TO-ANALOG CONVERSION)**   Converting continuously varying analog signals into their discrete digital counterparts.

**ANTENNA**   A resonant device that picks up faint radio signals and feeds them into a receiver.

**APERTURE**   The effective cross-sectional area of an antenna designed to transmit or receive electromagnetic signals.

**APOGEE (see also PERIGEE)**   The highest point along an elliptical orbit.

**ASCENDING NODE**   The specific point along the equatorial plane at which a satellite crosses the equator as it moves from the southern hemisphere to the northern hemisphere.

**ASTRODYNAMICS**   The practical application of celestial mechanics, propulsion theory, mathematics, and similar disciplines to the problem of planning and directing the trajectories of powered and coasting space vehicles.

**ATTENUATION**   (1) Any reduction in the strength or quality of an electromagnetic signal due to an intervening medium such as foliage, the ionosphere, or the atmosphere. The intervening medium distorts, reflects, and refracts the signal as it passes through. (2) The loss of power or signal quality of a modulated electromagnetic wave usually caused by intervening obstructions such as foliage or the charged particles in the ionosphere.

**ATTITUDE**   The orientation of a spacecraft with respect to an arbitrary set of reference axes.

**AVIONICS**   Any of the various electronic systems carried onboard an airplane.

**BANDWIDTH**   An adjacent span of frequencies occupied by an electromagnetic signal.

**BAUD**   The number of bits of information transmitted per second.

**BENT-PIPE COMMUNICATION SATELLITE (see also STORE-AND-FORWARD COMMUNICATION SATELLITE)**   A communication satellite that picks up signals from the ground and immediately rebroadcasts them on a different frequency toward a different geographical region.

**BIPROPELLANT ROCKET**   A rocket powered by two separate liquids: fuel, and oxidizer. The propellants are pumped or fed under pressure into the combustion chamber where they are mixed and burned.

**BIRD**   Industry slang for a communication satellite.

**BIT**   A single binary 1 or binary 0.

**BIT-ERROR RATE**   The number of errors, expressed as a fraction of the total number of bits sent usually in a digitally modulated channel.

**BROADCASTING**   The process of transmitting an electromagnetic signal to multiple, widely dispersed receivers.

**BYTE**   An adjacent sequence of binary digits, usually, but not always, eight in number. A byte is used in representing a single letter, number, special symbol, or punctuation mark.

**CARRIER WAVE**   A sinusoidal electromagnetic wave usually, but not always, modulated with useful information.

**CELLULAR TELEPHONE SYSTEM**   A ground-based mobile telephone system in which telephone calls are relayed through numerous low-power transmitters arranged in a gridlike pattern within an urban area.

**CHANNEL**   An adjacent band of frequencies that contain a specific broadcast signal.

**CHIPPING RATE**   The rate at which an electronic circuit produces binary digits.

**CIRCUIT**   A complete closed electrical pathway that controls the flow of electrons or other submicroscopic charged particles.

**CIRCULAR POLARIZATION (see also LINEAR POLARIZATION)**   The process of transmitting electromagnetic signals in a rotating corkscrewlike pattern.

**COAXIAL CABLE**   A pair of conductors consisting of a central conductor surrounded by an outer conductor. Coaxial cables have relatively good immunity to interference and low power losses when carrying high-frequency transmissions.

**CODEC**   The abbreviation for coder/decoder, a device used in digital communication systems to convert analog signals to digital signals, and vice versa.

**CODE DIVISION MULTIPLE ACCESS (see also FREQUENCY DIVISION MULTIPLE ACCESS and TIME DIVISION MULTIPLE ACCESS)**   A broadcast system in which each transponder spreads its modulated signal over the entire bandwidth assigned to all the common carriers.

**COMMUNICATION LINK**   Any physical or free-space connection between one location and another used for the purpose of transmitting and receiving information using electromagnetic waves as the carrier medium.

**COMMUNICATION SATELLITE**   An electronic relay station orbiting in space that picks up messages transmitted from the ground and retransmits them to a distant location, usually on a different frequency.

**COMSAT**   A specific satellite communications organization that represents America's interests in the Intelsat and Inmarsat consortiums. Also industry slang for a communication satellite.

**CONIC SECTIONS**   Circles, ellipses, parabolas, and hyperbolas created by passing a plane through a cone at various angles. To a first approximation, a spacecraft freely coasting in the gravitational field of a spherical celestial body such as the earth travels along one of the conic sections.

**CONSTELLATION**   Any collection of similar satellites designed to provide multiple coverage or multiple redundancy. Also, a group of stars regarded to be grouped together into a recognizable pattern by the people living on earth.

**DECIBEL**   A physical unit for expressing the ratio of two quantities of signal power. One decibel is equal to 10 times the common logarithm of this ratio.

**DEORBIT**   The process of firing a rocket onboard an orbiting space vehicle to reduce its orbital energy so it will reenter the earth's atmosphere and come back to earth either gently or catastrophically.

**DIGITAL COMPUTER (see also ANALOG COMPUTER)**   A computer whose processing functions are based on counting discrete entities rather than measuring continuous variables.

**DIGITAL MODULATION (see also ANALOG MODULATION)**   The process of converting incoming voice, video, or data signals into a form suitable for transmission over electromagnetic channels toward distant receivers.

**DIGITAL-TO-ANALOG CONVERSION (see also ANALOG-TO-DIGITAL CONVERSION)**   The process of converting discrete digital signals into their continuously varying digital counterparts.

**DOPPLER SHIFT**   Any systematic change in the frequency of a carrier wave that results when transmitter and receiver are moving at different velocities with a component of that velocity toward or away from one another.

**DOWNLINK (see also UPLINK)**   Any space-to-earth telecommunications pathway.

**DRAG**   The friction between a vehicle moving through the earth's atmosphere opposing the vehicle's forward motion.

**DUPLEX TRANSCEIVER (see also SIMPLEX TRANSCEIVER)**   A transceiver that can transmit and receive simultaneously so that the parties communicating can carry on a natural conversation.

**EARTH STATION**   Any ground-based communication facility capable of transmitting, receiving, and processing data relayed to and from orbiting satellites.

**ECCENTRICITY**   The "oblateness" of a satellite's orbit. The eccentricity of any orbit equals the difference between its apogee radius and its perigee radius divided by their sum.

**ECHO BALLOON**   One of the early spherical passive communication satellites 100 or 135 ft in diameter, constructed by coating the skin of a Mylar balloon with aluminum.

**ECHO SUPPRESSION**   The process of attenuating the echo effect in a satellite-based communication channel.

**ECLIPSE INTERVAL**   The period of time during which a satellite is shrouded in darkness within the earth's shadow.

**EFFECTIVE RADIATED POWER (ERP)**   The power, expressed in watts, that is radiated in the direction of the maximum antenna gain.

**ELECTROMAGNETIC WAVES**   Any sinusoidal carrier wave created by mutually orthogonal electric and magnetic fields that travels through a vacuum at an invariant speed of 186,000 mi/s.

**ELECTRON**   A subatomic particle with a specific negative charge and a very small mass that revolves around the nucleus of an atom.

**ELECTRONICS**   A specific branch of physics concerned primarily with the natural and controlled flow of electrons and other submicroscopic charged particles through various substances.

**ELEVATION ANGLE**   The angle between the local horizon and the line-of-sight vector pointing toward a satellite.

**ELLIPSE**   A closed, oval-shaped curve.

**EPHEMERIS**   Any complete set of numbers that, taken together, specify the approximate orbit of a satellite.

**FIBER OPTICS**   The technology of modulating information onto light beams transmitted through long, ultrathin strands of glass.

**FOLIAGE ATTENUATION**   Any reduction in signal strength or signal quality resulting from the limbs and leaves of trees that happen to be situated along the signal's line-of-sight path.

**FOOTPRINT**   A closed contour on the ground within which a communication satellite's beam provides sufficient signal strength to service customer needs.

**FREQUENCY (see also WAVELENGTH)**   The number of complete sinusoidal cycles an electromagnetic carrier wave goes through in 1 second.

**FREQUENCY DIVISION MULTIPLE ACCESS (see also CODE DIVISION MULTIPLE ACCESS and TIME DIVISION MULTIPLE ACCESS)**   Any broadcast system in which each transponder operates within its own assigned frequency slot or bandwidth.

**FREQUENCY REUSE**   Any of several techniques for maximizing the capacity of a ground-based or a space-based communication system. Frequency reuse is accomplished by isolating or polarizing the signals so carrier waves oscillating at the same frequency can service more users than would otherwise be possible.

**GEOCENTRIC**   Earth centered.

**GEODETIC**   Pertaining to the earth.

**GEOLOCATION (see also NAVIGATION)**   The process of fixing the position of a moving or stationary craft in longitude and latitude on the surface of the earth.

**GEOSYNCHRONOUS ORBIT (GEOSTATIONARY ORBIT)**   A circular orbit with a 24-hour period 19,300 nautical mi above the earth's equator. A satellite in a geostationary orbit appears to remain at a fixed location in the sky as seen by observers on the spinning earth.

**GIGAHERTZ**   One billion cycles per second.

**GRAVITY GRADIENT STABILIZATION**   A passive method of spacecraft stabilization and attitude control in which an elongated satellite (usually equipped with a telescoping boom) is oriented with its long axis pointing radially away from the earth so it can take advantage of the inverse square decrease in the earth's gravitational field to create torques that help keep it in its natural vertical orientation.

**GROUND TRACE**   The path a satellite traces out over the surface of the spinning earth as it coasts forward at its natural orbital speed around its circular or elliptical orbit.

**HERTZ**   One cycle per second.

**HOHMANN TRANSFER MANEUVER**   A two-burn-powered flight sequence that carries a satellite from one circular orbit to another along a specific transfer ellipse tangent to the two circular orbits at its perigee and apogee. The satellite traverses 180 degrees as it travels along the transfer ellipse between the two burns.

**HUB STATION**   The master station through which all communications to and from the users must flow.

**HYPERGOLIC PROPELLANTS**   Any fuel–oxidizer propellant combination that bursts into flame spontaneously on contact.

**INCLINATION ANGLE**   The angle between the orbital plane of a satellite and the equatorial plane of the earth.

**INMARSAT**   A specific international organization devoted primarily to providing telecommunication services for ships at sea.

**INTELSAT**   A specific international organization devoted primarily to providing land-based telecommunication services for a worldwide class of users.

**INTERFERENCE**   Randomly varying electromagnetic energy that tends to distort and attenuate the desired signals being picked up.

**IONOSPHERE**   Four roughly concentric layers of charged particles in the earth's upper atmosphere. The particles in the ionosphere bend, distort, and reflect electromagnetic waves.

**Ka-BAND**   The range of electromagnetic frequencies from 18 to 31 GHz.

**KILOHERTZ**   One thousand cycles per second.

**Ku-BAND**   The range of electromagnetic frequencies from 10.9 to 17 GHz.

**L-BAND**   The range of electromagnetic frequencies from 950 to 1450 MHz.

**LIGHT**   Those electromagnetic waves to which the human eye is sensitive.

**LINEAR POLARIZATION (see also CIRCULAR POLARIZATION)**   A process of transmitting electromagnetic signals such that they are made to oscillate parallel to one another, usually in the horizontal or vertical directions.

**MEGAHERTZ**   One million cycles per second.

**MICROSECOND**   One-millionth of a second.

**MILLISECOND**   One-thousandth of a second.

**MOBILE COMMUNICATION SATELLITES**   Any satellite that provides voice and data relay services between freely moving users on or near the ground below.

**MODEM**  An electronic device that modulates outgoing electromagnetic signals at the transmission end of a communications channel and/or demodulates the incoming signals at the receiving end.

**MODULATION (see also ANALOG MODULATION and DIGITAL MODULATION)**  (1) The process of manipulating the frequency or the amplitude of a carrier wave in response to an incoming voice, video, or data signal. (2) The process of encoding an electromagnetic carrier wave with useful information.

**MOMENTUM WHEEL**  A heavy flywheel rigidly mounted to the inside of a satellite whose rotational speed can be adjusted with the power from onboard electric motors to help maintain and adjust the spacecraft's angular orientation in real time.

**MONOPROPELLANT ROCKET**  A liquid rocket powered by a single liquid such as hydrazine. The monopropellant is sprayed over a catalyst to induce the desired chemical reaction, thus creating thrust.

**MULTIPATH**  A specific distortion that results when a portion of an electromagnetic signal is reflected from nearby surfaces, thus smearing and distorting the main signal which arrives by a more direct route. The so-called "ghosts" seen on a television screen are created by multipath reflections.

**MULTIPLEXING**  A specific body of electronic processing techniques that allow several simultaneous messages to be sent over a single communication channel at the same time.

**MULTISTAGE ROCKET**  An arrangement of rockets usually stacked one atop the other or mounted in parallel side-by-side. As the fuel in each stage is exhausted, the empty stage is jettisoned so the smaller rockets that remain will not have such a heavy load to push up toward orbital velocity.

**NANOSECOND**  One-billionth of a second.

**NAVIGATION (see also GEOLOCATION)**  The process of fixing the position of a craft and directing that craft along a preferred route from one known location to another.

**NOISE**  The unwanted and unmodulated electromagnetic energy that always corrupts, to some extent, the modulated signals being sent through a communication channel.

**OMNIDIRECTIONAL ANTENNA**  An antenna that transmits or receives electromagnetic signals to or from all directions.

**OPTICAL FIBERS**  Hair-thin cylinders of glass through which optical pulse trains are transmitted toward distant recipients.

**ORBIT**  The gravity-induced path followed by a satellite as it coasts around the earth or any other celestial body.

**ORBITAL MECHANICS**  A specific branch of physics and mathematics devoted to the analysis, prediction, construction, and modification of satellite orbits.

**ORBITAL PERIOD**  The amount of time it takes an artificial satellite to make one complete circuit around the earth.

**ORBITAL SLOT**  A longitudinal location along the geosynchronous arc specifically assigned to a particular satellite.

**ORBITAL VELOCITY**  The speed required for an artificial satellite to remain in orbit at a particular orbital altitude.

**PACKET SWITCHING**   A specific message-switching technique in which the messages traveling through a communication channel contain all the information necessary for routing, control, and self-checking for any coding and transmission errors.

**PARITY CHECKING**   An automatic computer-based error detection procedure that uses extra checking bits that are carried along with the usual numerical bits being processed.

**PARKING ORBIT**   Any temporary orbit, usually circular, within which a space vehicle awaits powered transfer to another orbit.

**PASSIVE COMMUNICATION SATELLITE**   A simple satellite with no moving or electronic parts that reflects electromagnetic waves from one ground station to another.

**PAYLOAD**   The electronic and mechanical devices carried onboard a space vehicle designed to help it perform its useful mission.

**PERIGEE (see also APOGEE)**   The lowest point along an elliptical orbit.

**PERTURBATION**   Any unwanted disturbance that distorts the regular orbital motion of a satellite.

**PHOTON**   An elementary bundle of electromagnetic radiant energy.

**PICOSECOND**   One-thousandth of a nanosecond, i.e., one-trillionth of a second.

**POLAR ORBIT**   Any orbit that carries an orbiting satellite alternately over the North and South Poles.

**PROPELLANT**   Any liquid, solid, or gaseous substance carried onboard a rocket or a satellite for the purpose of developing translational or attitude-control thrust.

**PROTON**   A subatomic particle with a positive charge, often located in the nucleus of an atom.

**RECEIVER (see also TRANSMITTER)**   Any electronic device that enables the desired modulated signal to be separated from all the other signals coming into the antenna.

**REENTRY**   The return of a spacecraft into the earth's atmosphere.

**REFRACTION**   The bending of an electromagnetic wave due to density variations in the medium (water, air, metal) through which that electromagnetic wave is passing.

**ROAMING**   A cellular telephone's usage of any cellular telephone switch other than its home switch.

**SATELLITE**   (1) Any artificial or natural body orbiting a larger body such as the earth. The satellite is held in its circular or elliptical trajectory by the combination of its tangential velocity and the gravitational force of the larger body pulling it radically inward toward the center of the gravitational field. (2) Any object that orbits the earth.

**SCRAMBLER**   A device that electronically alters a modulated electromagnetic signal so only those recipients equipped with special decoders can pick up and interpret a pristine version of the original signal.

**SEMICONDUCTOR**   Any substance such as silicon, germanium, or selenium that normally insulates against the flow of electricity, but that can, when infused with trace amounts of certain specific impurities, be made to conduct the flow.

**SIGNAL-TO-NOISE RATIO**   The power of the modulated signal being received, divided by the random electromagnetic noise coming in on that same channel.

**SIMPLEX TRANSCEIVER (see also DUPLEX TRANSCEIVER)**   A transceiver that is able to transmit or receive, but not both simultaneously. With a simplex transceiver the two parties communicating must talk and listen alternately.

**SOLAR ARRAY**   An adjacent set of solar cells used in converting incoming solar energy into electricity.

**SOLID-STATE DEVICE**   An electronic valve or amplifier composed of solid, monolithic materials whose electrical properties are controlled by specific impurities purposely infused into its crystal lattice structure during manufacture.

**SPACECRAFT**   Any vehicle designed to operate in space.

**SPECIFIC IMPULSE**   An abstract measure of the efficiency of a rocket propellant combination. The specific impulse equals the number of seconds a pound of the propellant will burn to produce a pound of thrust.

**SPECTRUM**   An adjacent set of electromagnetic waves.

**SPIN STABILIZATION (see also THREE-AXIS STABILIZATION)**   A specific form of satellite stabilization and altitude control in which the entire satellite spins at a fixed rate about a specific axis of spin.

**SPOT-BEAM**   A focused, conical-shaped pattern of electromagnetic waves directed from the antennas of a communication satellite toward a specific geographical region on the ground.

**SPREAD-SPECTRUM SIGNAL**   Any modulated signal superimposed on an electromagnetic wave with the characteristic that the number of bits of useful information being transmitted is appreciably less than the bandwidth of the transmission. Spread-spectrum transmission techniques provide improved jamming immunity, multipath rejection, and ranging accuracy for broadcast systems.

**STATIONKEEPING MANEUVERS**   On-orbit rocket burns used in making small orbital adjustments to keep an orbiting satellite within an acceptable distance of its assigned orbital slot.

**STORE-AND-FORWARD COMMUNICATION SATELLITE (see also BENT-PIPE COMMUNICATION SATELLITE)**   A communication satellite that picks up signals from the ground, records them on electronic or magnetic storage media, then rebroadcasts them on command when it is over some distant geographical region.

**SUBLIMATION**   The process by which a solid substance changes directly into a gas without passing through the intervening liquid state.

**TELEPROCESSING**   The use of telephone lines or other communication channels to transmit useful information between remote locations.

**THREE-AXIS STABILIZATION (see also SPIN STABILIZATION)**   A specific form of satellite stabilization and control in which the satellite's main body is made to maintain a fixed attitude relative to the earth's surface and its orbital track.

**THRUSTER**   A small onboard rocket used by a satellite for stationkeeping maneuvers, momentum dumping, and altitude control.

**TIME DIVISION MULTIPLE ACCESS (see also FREQUENCY DIVISION MULTIPLE ACCESS and CODE DIVISION MULTIPLE ACCESS)**   A broadcast system in which each transponder operates within its own assigned discrete time slots.

**TRANSCEIVER**   Any electronic device that can transmit and receive modulated electromagnetic waves.

**TRANSFER ORBIT**   An intermediate elliptical orbit along which a satellite travels from its initial low-altitude orbit out to a higher-altitude orbit.

**TRANSMITTER (see also RECEIVER)**   Any electronic device that broadcasts modulated electromagnetic signals toward one or more distant receivers.

**TRANSPONDER**   An electronic device carried onboard a communication satellite that picks up signals from the ground on one frequency and immediately rebroadcasts them on a different frequency.

**ULTRAHIGH FREQUENCY (UHF)**   That specific band of frequencies ranging from 300 to 3000 MHz.

**UNIVERSAL LAW OF GRAVITATION**   A specific physical law discovered by Isaac Newton which states that all material bodies in the universe attract one another with a force that is directly proportional to the product of their masses and inversely proportional to the distance between their centers.

**UPLINK (see also DOWNLINK)**   Any earth-to-space telecommunications pathway.

**VAN ALLEN RADIATION BELTS**   Two large donut-shaped rings of high-energy charged particles (electrons and protons) spiraling around the earth's magnetic lines of flux.

**VELOCITY**   The speed at which an object moves in a particular direction.

**VERY HIGH FREQUENCY (VHF)**   That specific band of frequencies ranging from 30 to 300 MHz.

**VOLATILE STORAGE**   A computer storage medium in which the information being stored vanishes if the power is turned off or temporarily interrupted.

**WAVELENGTH**   The distance between two adjacent crests or troughs of a wave.

**ZENITH**   The point in the sky directly above an observer who is located on the ground or in space.

# Historical Perspectives

*In the year 2054, the entire defense budget will purchase just one tactical aircraft. This aircraft will have to be shared by the Air Force and the Navy 3 $^1/_2$ days each per week except for leap year, when it will be made available to the Marines for the entire day.*

    *Playful prediction made by Norman R. Augustine, president of Martin Denver, after he made a graph showing the relentless exponential growth rates in the cost of and complexity of military airplanes, 1983*

**1609**   German astronomer Johannes Kepler publishes his first two laws of planetary motion dealing with orbit shape and orbital speed. Ten years later, in 1619, he supplements his first two laws with a third law linking a planet's average radial distance with its orbital period.

**1686**   Englishman Isaac Newton publishes his landmark work *Principia,* unifying virtually all the physics and mathematics developed up to his day.

**1801**   Frenchman Joseph Jacquard utilizes the first known punched cards to control his complicated weaving looms.

**1835**   Portrait painter Samuel Morse develops his famous binary Morse code made up of "dots" and "dashes." Three years later, he uses the new code to transmit messages through short telegraph lines.

**1844**   Samuel Morse transmits a coded telegraph message through telegraph wires strung across a distance of 41 mi between Washington, D.C. and a railway station in Baltimore, Maryland. The message says: "What hath God wrought?"

**1854**   English logician and mathematician George Boole publishes his first book on symbolic logic, later to form the backbone of formal computer design.

**1864**   James Clerk Maxwell at London University proves mathematically that electromagnetic waves invisible to the naked eye should exist.

**1866**   The first successful transoceanic telegraph cable is laid under the Atlantic Ocean.

**1869**   American clergyman and author Edward Hale publishes a widely circulated short story entitled "The Brick Moon," in which he describes an artificial satellite launched into space to serve as an artificial navigation aid.

**1876**   Alexander Graham Bell summons his assistant over one of the world's earliest telephones, "Mr. Watson, come at once. I want you." Bell needed immediate help because he had spilled acid down the front of his trousers. Nearly all subsequent telephone calls were less important than that first call.

1878    The first commercial telephone exchange in the United States is opened for service in New Haven, Connecticut. It provided eight lines and served 21 telephones.

1886    Heinrich Hertz first uses oscillatory radiating circuits to demonstrate the production of free-space electromagnetic waves identical to the ones that were predicted mathematically by James Clerk Maxwell 22 years earlier.

1888    The first frequency division multiplexing systems are placed in service to allow multiple telephone conversations and telegraph messages to be sent through the same wires at the same time.

1901    Guglielmo Marconi successfully picks up a wireless coded signal at Signal Hill, Newfoundland, which was sent across the Atlantic from Polhu, Cornwall in the English countryside.

1905    The thermionic emission phenomenon, first discovered by Thomas Edison in 1863, is harnessed by Englishman J. A. Fleming in the construction of the first pickle-sized vacuum tube.

1908    American inventor Lee DeForest uses his powerful intellect to develop and perfect the vacuum triode, a clever electronic amplification device.

1918    Massachusetts college professor Robert Goddard conducts a successful flight test of his first solid-fueled rocket.

1929    Robert Goddard launches his first (extremely noisy) liquid-fueled rocket from his Aunt Effie's cabbage patch in Auburn, Massachusetts.

1929    Austrian engineer N. Noordung publishes a technical paper advancing the theory of geostationary earth-orbiting satellites.

1940    George R. Stibitz demonstrates the first transmission of computer-generated digital pulse trains over long-distance telephone lines.

1940s    AT&T engineers propose, but do not develop, the first cellular telephone system using multiple low-power transmitters scattered around the service area and rigged for extensive frequency reuse.

1945    Arthur C. Clark describes the use of geosynchronous communication satellites to serve the communication needs of future earthlings.

1946    After touring the data processing facilities at the Moore School of Engineering, the brilliant Hungarian immigrant, John Von Neumann publishes his 40-page "blue book" lucidly describing his incisive ideas on future computer design.

1948    John Bardeen, Walter Brattain, and William Shockley patent the first practical solid-state device, the point-contact transistor.

1955    John R. Pierce at Bell Laboratories publishes a technical paper laying out, with surprising accuracy, the technical requirements for a satellite-based communication system.

1956    The first transoceanic telephone cable is successfully laid under the Atlantic Ocean.

1957    Soviet aerospace engineers launch the first Sputnik into space, thus awakening the world to the vast potential for exploiting the special environmental properties found along the space frontier.

1958    Aerospace scientists launch America's first artificial satellite, *Explorer I,* which is used in detecting the Van Allen Radiation Belts curling around the earth.

1958    America's *Score* satellite relays a prerecorded message from President Eisenhower to the citizens of planet Earth.

1960    The first Echo balloon, a 100-ft aluminized Mylar passive communication satellite, begins to reflect electromagnetic waves directed up toward it from the ground.

1960    The first successful active communication satellite, *Courier 1,* begins rebroadcasting 2-GHz electromagnetic signals back from space.

1960    The world's first practical navigation satellites, called Transit, are successfully launched into low-altitude polar orbits.

1961    Yuri Gagarin of the Soviet Union becomes the first earthling to orbit the earth.

1963    AT&T researchers design and launch the *Telstar* communication satellite, a talky electronic beach ball.

1963    More than 1 billion 3/4-in. copper dipole "needles" are launched into space under the code name "Project West Ford." As expected, they form an "artificial" ionosphere capable of bouncing 8-GHz signals back from space.

1964    *Syncom II,* the first geosynchronous communication satellite, arrives at its 19,300-nautical mi orbital destination.

1964    The first 19 government representatives sign the international Intelsat agreement. Today that organization boasts more than 120 member states.

1965    American researchers R.C. Platzek and J.S. Kilby first elucidate a startlingly original concept for developing the first solid-state integrated circuit.

1965    The first of the Soviet Molniya communication satellites begins blanketing the Soviet empire with radio and television programs from its 12-hour elliptical orbit tipped at an angle of 63.4 degrees.

1972    Microprocessor chips are first marketed in large commercial quantities.

1972    The Federal Communications Commission (FCC) announces America's "Open Skies" policy whereby any financially sound organization serving the public interest can own and operate its own satellite.

1972    The first Landsat satellite begins to inventory the valuable resources and the land-use patterns on the ground below.

1974    The first privately owned domestic communication satellite, *Westar-I,* is launched into orbit to serve the commercial needs of Western Union.

1975    The first terrestrial cable television system operated by HBO (Home Box Office) initiates regular satellite-relay operations.

1976    America's first nuclear-powered communication satellite begins broadcast operations from its orbit in space.

1976    The first Marisat satellite is launched into space to provide shipboard telecommunication services.

1977    Rockwell International's 3-year space industrialization study ends with a rather heavy 1100-page summary report. One section of the report includes detailed plans for constructing a gigantic orbital antenna farm rigged to provide five practical communication capabilities including the servicing of a worldwide collection of hand-held cellular telephones.

1979   The international Inmarsat organization serving maritime nations draws 26 initial signatories. Today it has about 60 member states.

1979   The Soviet Union's first geosynchronous communication satellite reaches its orbital destination and begins broadcast operations as a part of the Soviet empire's Intersputnik communication system.

1979   The first cellular telephone system is set up in Tokyo, Japan. It has a capacity for 4000 subscribers with capabilities for easy expansion to 8000.

1981   The maiden flight of space shuttle *Columbia* takes place from a Cape Canaveral launch site. Later in that same year, a refurbished *Columbia* takes to the skies with its first useful payload.

1981   *Time* magazine names the digital computer "Man of the Year."

1983   American astronauts release the first Tracking and Data Relay Satellite (TDRS) from the open cargo bay of the space shuttle. The 5000-lb TDRS satellite is capable of transmitting down to earth the entire contents of a 24-volume encyclopedia in just 5 seconds.

1984   The space shuttle astronauts establish a space first by recovering a disabled satellite and returning it to earth for refurbishment and relaunch.

1986   Seven American astronauts tragically lose their lives in the shuttle *Challenger* spacebound flight.

1987   The USSR's huge heavy-lift booster, Energiya, is successfully hurled into orbit from the Baikonar cosmodrome.

1988   Edward Tuck develops the basic concepts for the 840-satellite Teledesic constellation. Tuck, who earlier founded the Magellan Company, the world's second largest Navstar user-set maker, later told a reporter how the unique navigation capabilities of the Navstar triggered the basic idea for Teledesic. "I thought, if I go to the North Pole, I have something that tells me I am at the North Pole," he explained. "Wouldn't it be nice to pick up a phone and call someone and say, 'I'm at the North Pole.'"

1990   The Hubble space telescope is released in a low-altitude orbit by a team of space shuttle astronauts.

1992   Russian researchers begin testing a new commercial communication satellite called Gonets. Gonets is designed to serve 5-lb transportable telephones.

1993   American and Russian negotiators reach an accord whereby some American satellites will be carried into space aboard launch vehicles built and tested by aerospace engineers in the former Soviet Union.

1993   Motorola successfully raises $800 million to begin building the satellites that will populate its 66-satellite Iridium constellation.

1993   In a short flight test, the DC-X Delta Clipper unmanned single-stage rocket successfully ascends vertically over a New Mexico launch pad, then descends vertically for a soft landing on the desert sand.

1994   Advanced gallium–arsenide computer chips are introduced into AT&T's new generation of cordless phones.

1994   Billionaires Bill Gates and Craig McCaw join forces with entrepreneur Edward Tuck, founder of Calling Communications, to form a new company, Teledesic, for the purpose of launching an 840-satellite mobile communication system into space.

1994   Loral officials make an official announcement via a five-city satellite hookup that they have succeeded in raising $275 million in equity financing for their Globalstar constellation.

1994   Russian aerospace engineers begin to publicize the benefits of their 42,000-lb Globis orbital antenna farm. They intend to send Globis into a geosynchronous orbit aboard their Energiya booster rocket.

1994   The controlling interest in Craig McCaw's cellular telephone company is acquired by AT&T for an estimated $12.6 billion in AT&T stock.

1994   The Navstar Global Positioning satellite system's 24th Block II satellite begins to serve customers, thus completing the operational Navstar constellation.

## The Future?

1994   The first two Orbcomm communication satellites developed by researchers at the Orbital Science Corporation begin demonstration service as a precursor to their 26-satellite mobile constellation.

1995   Inmarsat reaches a user base of 100,000 customers worldwide.

1996   The first Iridium satellite in Motorola's $3.4 billion constellation arrives in its 413-nautical mi polar orbit.

1997   Portable Satcom telephones are widely available for less than $1000 each.

1998   All 66 of Motorola's Iridium satellites are now in space.

2000   As a KPMG Peak Marwick study had indicated in 1994, total revenues from satellite-delivered mobile telephone services exceeded $11 billion this year. Although several new systems are on the verge of coming online, the more optimistic high-end forecasts for $22 billion in revenues does not materialize.

2001   The Teledesic constellation consisting of 840 low-altitude communication satellites reaches operational status in the closing months of this widely celebrated year.

2002   The space shuttle astronauts fly a special mission to service the Hubble telescope in its 330-nautical-mi-high orbit.

2005   Inmarsat reaches a customer base of 1 million users worldwide.

2054   The Department of Defense expends the entire defense budget on a single tactical airplane. The Air Force and the Navy are each allowed to share it 3½ days each week, except for leap year, when it is made available to the Marines for the entire day.

# Index

Abyss, the, 201
acceleration, 255
acoustic coupler, 28–29, 255
action reaction principle, 53
active communication satellites, 42
Acunet USA, 39
air traffic control, 118, 120–123
aircraft landing and taxiing, 122–125
Airfone, 34, 112–114, 116, 171
Alcorn, Fred, 166
alphanumeric messaging, 28
altitude and velocity control subsystem, 52
altitude control, 52–53
altitude-dependent characteristics, 130
American Mobile Satellite Corporation
      (AMSC), 150
amplitude modulation (AM), 17–19
AMPS (Advanced Mobile Telephone System),
      15–16
analog cellular telephones, 188
analog computer, 255
analog modulation, 19, 255
analog-to-digital conversion, 255
antenna, 255
antenna design, 151
aperture, 255
apogee, 255
Apollo 13 mission, 228
Archimedes constellation, 210–212
Archimedes screw, 210
Argos messaging system, 73, 75–77, 79–80
Aries constellation, 193
Aries satellites, 194
armchair vacations, 203
artificial satellite, 111
ascending node, 255
asparagus stalk booster, 234, 236
Astaire, Fred, 230
astrodynamics, 255
atomic clocks, 18, 89
atomic oxygen, 133
ATS-6 satellite, 46
attenuation, 255
attitude, 256
Augustine, Norman, R., 265

Australia's Optus Communications, 157
automotive navigation, 102
avionics, 256

bag phone, 25
ballistic parameter, 59
bandwidth, 256
bandwidth compression, 64
Bardeen, John, 266
base station, 13–14, 20
battery life, 23
baud, 256
beam-making machines, 65
beeping grapefruit, 37, 102
Bell Canada, 156
beneficial properties of outer space, 38
bent-pipe communication satellite, 256
bent-pipe message relay, 74, 191, 205
bent-pipe radionavigation, 75
beta tin, 41
Big Wind of 1938, 211
binary pulse trains, 32
bipropellant rocket, 256
bird, 256
bit, 256
bit error rate, 256
board meeting in outer space, 166
Boole, George, 265
booster rocket, 49, 51
Brattain, Walter, 266
broadcasting, 256
brontosauruslike creatures, 120
byte, 256

Calling Communications, 196, 269
Canada's Aurora 400 system, 17
Captain Kirk, 26
car phone, 21
cargo shipping containers, 108–109
carrier wave, 256
carrier-aided solutions, 124–126
cell-splitting techniques, 20
Cellular Digital Packet Data (CDPD), 28–29

271

## ABOUT THE AUTHOR

Tom Logsdon is a senior member of the technical staff at Rockwell International in Seal Beach, California. An award-winning author, lecturer, mathematician, and aerospace engineer, he has performed systems analysis and computer simulations for the Saturn V moon rocket project, Skylab, and NASA's reusable space shuttle. Mr. Logsdon is the author of 25 books on space technology, computer science, and robotics, including the recently published *The Navstar Global Positioning System*. He is also world-renowned for his short-course seminars on space technology.